The Interstellar Age

ALSO BY JIM BELL

Postcards from Mars
Mars 3-D
Moon 3-D
The Space Book

The

INTERSTELLAR AGE

Inside the Forty-Year
Voyager Mission

JIM BELL

DUTTON
— est. 1852 —

DUTTON
→ est. 1852 ←

Published by the Penguin Group
Penguin Group (USA) LLC
375 Hudson Street
New York, New York 10014

USA | Canada | UK | Ireland | Australia | New Zealand | India | South Africa | China
penguin.com
A Penguin Random House Company

LIBRARY OF CONGRESS CATALOGING-IN-PUBLICATION DATA
has been applied for.

ISBN 978-0-525-95432-3

Printed in the United States of America
1   3   5   7   9   10   8   6   4   2

Designed by Katy Riegel

To the men and women of Voyager,
and the generations that they have inspired

Contents

Part Three

LOOKING BACK, LOOKING AHEAD

The Interstellar Age

Prelude: Outbound

I believe our future depends, powerfully, on how well we understand this Cosmos in which we float like a mote of dust in the morning sky. We're about to begin a journey through the Cosmos . . . it's a story about us . . . how the Cosmos has shaped our evolution and our culture, and what our fate may be.

—Carl Sagan
(*Cosmos: A Personal Journey*)

PHYSICS TELLS US that all things attract each other gravitationally, from pulsars to planets to petunias, even if those forces are sometimes too small to notice in everyday life. But if you look closely at the trajectories that your life has taken, you may notice the results of similar gravitational effects from the people you have known. Sometimes people around us cause massive swings in direction and speed that can propel us on toward new and undiscovered territory and experiences. That's what happened with me and the space-exploration mission known as *Voyager*.

The trajectory of my life has been guided by the slow, gentle, persistent gravitational pull of two elegant robotic spacecraft and the teams of people—scientists, engineers, mentors, students—who made their missions of exploration so marvelously compelling. Taking advantage of a rare celestial alignment of the planets, those two robots, *Voyager 1* and *Voyager 2*, gave us all our first detailed, high-resolution, glorious views of the solar system beyond Mars, revealing the giant planets Jupiter, Saturn, Uranus, and Neptune, and their panoply of rings and moons, in all their awesome wonder—not just for scientists, but also for poets, musicians, painters, novelists, moviemakers, historians, and even kids.

I happened to have been born at a time that placed me in college and graduate school right when the fruits of that fortuitous celestial alignment were ripening. By a random turn of a corner in a building, while walking back from class, I spotted a flyer from a professor who was looking for student research help. I soon found myself *involved* in the missions of these extraordinary projections of human technology—something I had dreamed of since I could barely read. I felt as if I had been cast out into deep space myself, seeing my life, and my world, from a completely new perspective. In one seemingly chance Forrest Gump–like encounter after another, the arc of my life has been shaped by the *Voyager* missions, and even to this day I find myself drawn to their power to lift the human spirit. Just think of these sophisticated creations—mere machines, yet projections of ourselves—launched into my hero Carl Sagan's "shallow depths of the cosmic ocean," representing the integrated abilities, hopes, dreams, and fears of the more than 100 billion people who have lived on planet Earth and who, like me, have wondered, "Are we alone?" "What else is out there?" "What is our destiny?"

These Voyagers—and by that I mean the people as well as the machines—have taken us all on a tour of the Greatest Hits of the

Solar System, and we have all been privileged passengers carried along for the ride. Along the way, I went from a starry-eyed kid interested in astronomy and planetary science to a student learning the ropes from some of the greatest masters in the field, to—now— a practitioner of the art myself, with students of my own. It has been an adventure filled with astounding beauty, discovering new worlds so exotic that their alien landscapes were entirely unanticipated, facing unprecedented challenges, meeting and then saying goodbye to new friends and colleagues. . . .

And now the *Voyagers* are leaving the protective bubble of our sun and crossing over into the uncharted territory between the stars. They—and we, through them—are now interstellar travelers. Via their technology, their discoveries, and the messages that they are delivering to the galaxy on our behalf, we have all entered the Interstellar Age. This may be the ultimate legacy of the men and women and machines of *Voyager*. As we learn and grow as a species, as we begin to grasp the fragility of our existence and the fleeting nature of habitable environments in our solar system, we must adapt and move on. In the long run—the *very* long run—we will have to leave our sun's cradle and move out into the stars. The Interstellar Age is the inevitable future of humankind, and the *Voyagers* are our first baby steps along that path.

I want to share that story with you here and convey, I hope, how special it has been to be witness to what historians of the future will no doubt regard as some of the most incredible voyages of exploration ever attempted.

Part One

ALIGNMENT

1

Voyagers

I GAZED IN wonder at the graceful and swirling azure clouds of Neptune. I had impulsively boarded the spacecraft in 1977, at the age of twelve, seeking a place where the gravity of my world would be but a distant memory. Word of the launch came on the evening news. "A Grand Tour of the Solar System!" the announcer proclaimed. Two launches would carry our band of travelers destined for Jupiter and Saturn, and if all went well, we would forge on, perhaps past Uranus and Neptune—worlds as yet unexplored. The thought of running away from home to explore some distant land tugged at me, as it did for many preteens. In the world of my small Rhode Island town, even traveling to another area code may as well have been like traveling to Mars. Of course, it could only ever be a dream:

traveling faster than any rocket had ever gone, taking two years to Jupiter, three years to Saturn, Uranus by the mid-'80s, Neptune by my twenty-fourth birthday . . .

It sounds like science fiction, but this is essentially a true story. The spaceships are called *Voyager 1* and *Voyager 2*, and they really did launch in 1977. While the *Voyagers* don't carry humans on board, they do carry our eyes and ears, our most sophisticated cognitive intelligence, our science and art and dreams. In 1977 they brought me out of that sheltered world of childhood and into a fantastic new world of learning, culture, and Big Science, first as a college student at Caltech in Pasadena, then as a graduate student in Hawaii. *Voyager*'s story of exploration parallels my own. Indeed, the missions have touched countless lives and careers in space science and engineering—so many of the people I know and have worked with over the decades feel as if the *Voyagers* propelled their lives.

The "Grand Tour" announced that day in 1977 would take advantage of a once-every-176-year planetary alignment that provided an opportunity to send a single spacecraft past *all four* giant outer solar system planets, using the gravity of one to slingshot the mission on a path to the next, bouncing it from one remarkable world to another, and then eventually completely out of our solar system. The last time such an alignment occurred, back in the eighteenth century, the frontier of exploration was defined by European wooden sailing ships.

The twelve-year-old me had become hopelessly hooked on space exploration watching the adventures of the *Apollo* astronauts on the moon. My parents tell me that they woke me up on that Sunday night in July of 1969 to witness Neil Armstrong and Buzz Aldrin make history on live TV in the Sea of Tranquility. We saved the

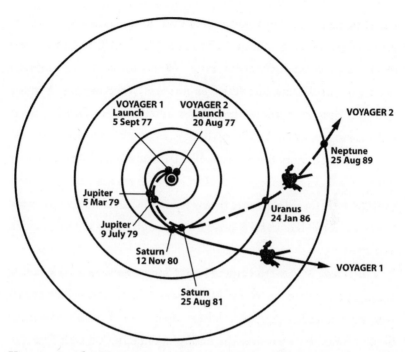

***Voyager 1* and 2 Trajectories.** Schematic diagram of the trajectories that enabled NASA's twin *Voyager* spacecraft to tour the four gas giant planets and achieve the velocity to escape our solar system. *(NASA/JPL)*

Monday MAN ON MOON! giant headlined copy of the Providence *Evening Bulletin*, which I later had framed. For the next three and a half years, I was glued to the television, whenever possible, watching these guys walking—and *driving cars!*—on the lunar surface. While I was assured by the voices of NASA engineers and space commentators that it was hard work, many of the astronauts seemed like they were having fun. I want to do *that*, I thought. I dressed as an astronaut for a long run of Halloweens.

I followed the exploits of the twin *Viking* landers sent to the surface of Mars in 1976. Even though people weren't going, the idea of sending two car-sized robots on a 150-million-mile remote-control journey and getting them to set down, softly, onto the surface of the Red Planet was astounding. In the decades ahead I would witness firsthand even crazier Mars landing systems as the *Mars Pathfinder* and *Spirit* and *Opportunity* rover mission set down on Mars—successfully—using bouncing airbags, and the larger *Curiosity* rover did so using its Rube Goldberg–like "sky crane" landing system. *Viking* used old-school technology, like parachutes and retro-rockets, right out of a Bugs Bunny cartoon. While Marvin the Martian wasn't waiting there for us, the Mars that was revealed by the *Vikings* turned out to be eerily like deserts on Earth, though much dustier, colder, and drier.

The early 1970s-era cameras on *Viking* were essentially faxing their photos back to Earth, and NASA was using what was then brand-new electronic imaging technology that needed no photographic film. Instead, it converted the sunlight reflected off of a Martian scene into radio signals, beaming them back to Earth, where the faint signals were picked up by radio telescopes the size of a baseball field. I saw the digital images these faint signals produced revealed on the nightly news. The first images came down live, and painstakingly slow—one column of picture elements or "pixels" at a time. Space photography! I want to do *that*, too, I thought. My parents and grandparents helped me buy a telescope and some attachments to link it up to my 35mm camera.

These days it's hard to explain to my kids, or to my students, what it was like growing up thirsty for science in the 1970s and 1980s. Imagine a world, I implore them, where there are only three major TV networks plus another run by the government, called the Public Broadcasting Service, or PBS. Imagine further that for the most part only the government channel would have science shows on TV (not counting *Star Trek*—one of my favorites, to be sure, but only partly "science"). For the most part, science TV was dominated at the time by *NOVA*—the educational and beautifully produced show from Boston's WGBH station that is still running strong today. But that was basically it: no Science Channel, Discovery Channel, National Geographic Channel, NASA TV, History Channel, or for that matter, no Fox, CNN, MTV, VCRs, DVRs, and *no way to skip the commercials.* They look at me in horror, as if I had to endure being raised by wolves in the frozen tundra. Then they shoot me a truly pitiful look when I remind them that, worse yet, we had no Internet. Gads! How did we survive?!

In that bleak landscape of science communication was the TV show *Cosmos*, which first aired on PBS in 1980. The show's host, the astronomer, planetary scientist, astrobiologist, *Voyager* imaging team member, and science popularizer Carl Sagan, was probably the first scientist I had ever encountered who spoke English. I mean common English, more like what you'd hear around the dinner table than the jargon and shorthand codes that most scientists typically use when talking about their work. But that plain talk was also laced with metaphor and analogy and evocatively grand cadences, often accompanied by the soaring and romantic electronic music of Vangelis. Sagan revealed the mysteries of the planets and moons and asteroids and comets and stars and galaxies and where we came from and where we're going. I found myself *listening to him* and falling in love with the idea of doing science, of possibly even becoming a scientist. It was a captivating, mind-blowing, entertaining glimpse into the modern world of astronomy and space exploration. I would eagerly await each week's new episode, talking about it endlessly the next day with my nerd friends at school, mimicking Sagan's distinct, guttural staccato voice . . . "Perhaps, one day, we will sail among the stars on gossamer beams of light. . . ." My mother loved his turtleneck and tweed jacket (I would get to introduce her to him many years later, at a professional conference we both attended in Rhode Island. We were both just starstruck, but Carl was kind, warm, and thoroughly approachable).

Back in 1977, it was clear to even a teenager that *Voyager* would be something very different from *Viking*. First of all, it would embark on a long journey. It would sail on for more than a decade at least, and if nothing bad happened along the way, the plutonium-fueled nuclear power pack that generated electricity for the spacecraft could keep the systems working for perhaps *fifty years*. With that kind of longevity, it was possible for the spacecraft to survive long enough to cross into interstellar space—the realm outside of our sun's protective magnetic cocoon. *Voyager* would then venture out into the strange and unfamiliar interstellar wind. Wild. And second, the mission had the potential to literally discover entirely new and alien worlds! *Viking* had made important discoveries on Mars, but the landscape and processes had generally been familiar: wind, sand, maybe a little water long ago, grinding down and carving into the rock, eroding landscapes like you might encounter on a car trip through northern Arizona, Utah, and southern Colorado. *Voyager* would be encountering worlds not of rock, but of ice and gas, places where the sun is only the blip of a flashlight in an otherwise black, starry sky, and where the temperature might be only a few tens of degrees above absolute zero.

Through my youthful eyes, the biggest appeal of *Voyager* was indeed this idea of exploring the truly unknown—throwing a bottle, of sorts, into the cosmic ocean and seeing where the eddies and currents of nature would take it. In my telescope, on a clear cold night, I could make out the reddish-brown belts and bands of Jupiter, as well as its famous Great Red Spot. It was a good-sized instrument for a young amateur astronomer, a so-called Newtonian telescope (designed by Isaac Newton, and using mirrors instead of lenses) made by a company called Meade Instruments, with a main mirror about eight inches in diameter and a tube about four feet long. With that tube held by a metal mounting post and three wide metal legs, it was a heavy, bulky, cumbersome thing to schlep outside and in from the garage and to set up every time I wanted to use it (especially in the snow), but it was so worth the effort. I could resolve the enchanting, creamy yellow rings of Saturn and learned firsthand why that planet was called "the jewel of the solar system" by the pioneers of astronomy. It always amazed me, in fact, that when I looked at Saturn I was seeing *the real Saturn*. Like a lot of kids at the time, I collected coins and stamps and baseball cards, and that was great, but there was always someone with an older, better, or cooler collection than mine. No one had a *better Saturn* in their planet collection—I was looking right at the real thing with my own eyes. I felt an unfamiliar sensation, an extraordinary lightness, as I took in the nearness of a world so distant.

In my small telescope, Uranus and Neptune, and the six or seven little moons I could sometimes find perched around Jupiter and Saturn, were just specks of light. It was hard to imagine these dots as worlds, as destinations one might visit, as lands of rock and ice, wind and volcanoes, polar caps, and panoramas so staggeringly familiar and yet patently alien. They were just dots, to me and everyone else on our little blue planet—even the largest telescopes in the world at the time couldn't reveal their true nature. But I knew that

Voyager would change that, that these dots would soon permanently become distinct places, as diverse in character as the myriad environments of our own planet, and far more exotic than our own moon (which had also, only recently, become a bona fide place, rather than a two-dimensional icon in the sky). The ability to ride along with *Voyager*, to be a passenger on this trailblazing journey destined to discover entirely new worlds, to *see history made*, was irresistible to the young me. In fact, it still is.

EXPEDITION LEADERS

Famous ships of exploration are usually led by a famous captain or commander, like Christopher Columbus, Ferdinand Magellan, James Cook, Ernest Shackleton, or Neil Armstrong. The *Voyagers*, however, are led by a committee of captains—managers, engineers, and scientists from NASA's Jet Propulsion Laboratory (JPL) and elsewhere who were tasked with overseeing the design, manufacture, and operation of the most ambitious robotic planetary exploration mission yet attempted—and a pair of equally powerful commanders, a project manager and a project scientist.

In NASA and JPL parlance, *Voyager* is a "Project" (capital *P*), run by a Project Office (capital *O*) and divided organizationally into a number of subsidiary offices. These include the Mission Planning Office, where the detailed spacecraft trajectories were designed; the Flight Science Office, which includes the science team and which is responsible for making sure the mission achieves its scientific objectives; the Flight Engineering Office, with the engineers and managers who designed and built *Voyager*'s power, thermal control, communications, and propulsion modules; the Flight Operations Office, which provides the procedures and software needed to plan and actually operate the spacecraft and its science instruments (and which includes two teams of JPL scientists and engineers: the spacecraft operations team, who directly communicate with the spacecraft and who monitor its status and health over time, and the science support team, who serve as the interface between the science team and spacecraft operations team); and the Ground Data Systems Office, which provides the hardware and software needed to send commands up to the spacecraft ("uplink") as well as to receive and process data back down from the spacecraft ("downlink").

The project manager leads the Project Office and is the engineering and management commander of *Voyager*, responsible for getting the spacecraft built and tested, keeping the mission safely operating on time and on budget, and overseeing the hundreds of contractors and several thousand engineers, technicians, and other managers on the Project. The project scientist runs the Flight Science Office and the science team—a group of scientists, engineers, technicians, managers, and students from around the world who designed, built, and operate the science instruments and who interpret the downlinked data. The project scientist is the scientific commander of *Voyager*, responsible for making sure the mission achieves its science goals on time and on budget and for coordinating and herding (like cats) the hundreds of scientists on the Project.

Each *Voyager* carries scientific instruments for eleven investigations. These include wide-angle and high-resolution cameras for imaging and spacecraft navigation; radio systems for studying gravitational fields and planetary radio emissions; infrared and ultra-violet spectrometers to measure chemical compositions; a polarization sensor for surface, atmosphere, and planetary ring composition; a magnetometer measuring magnetic fields; and four devices for studying charged particles, cosmic rays, plasma (hot ionized gases), and plasma waves. Scientists conducting each investigation are organized into instrument teams, and the leader of each instrument team is called the principal investigator (or PI). The PIs are responsible for the design, construction, and operation of each of their instruments, and together they form the Science Steering Group, which is chaired by the project scientist and which reports to the project manager.

In this kind of committee-led project, it's critical that the two commanders at the top of the org chart, the project manager and the project scientist, are consistently on the same page and have an excellent working relationship. Each is personally responsible for the success of the mission—to NASA and, ultimately, to Congress and the taxpayers who are footing the bill. Over the course of more than four decades since the Project began, *Voyager* has had ten project managers. But during that entire time, the mission has had *only one* project scientist: Edward C. Stone.

Ed Stone is a space weatherman, a physicist who studies the ways that high-energy particles called cosmic rays travel through space and interact with the magnetic fields and atmospheres of the sun and planets. Cosmic rays are a form of high-energy radiation made of protons and the nuclei of atoms, and they travel through the universe at nearly the speed of light. Exactly where they come from is still a mystery—they could be caused by massive supernova explosions of dying stars, or by the powerful black holes in the centers of active galaxies, for example. Regardless of how they formed, scientists like Ed can use the properties of cosmic rays to understand the details of the ebb and flow of the solar wind (the stream of high-energy particles coming off the sun) and the way that wind is carried by the sun's magnetic field and interacts with the magnetic fields of the planets. Measurements of this kind of "space weather" were some of the first scientific measurements ever made from space satellites, and Ed Stone has been a prolific scientist in this game since the beginning.

In 1972, Ed was appointed as the project scientist for *Voyager*. Over the course of the mission he's had other important roles as well, including serving as the director of JPL from 1991 to 2001 and as the PI for *Voyager*'s Cosmic Ray Subsystem (CRS) instrument, which is making the measurements that are most closely aligned to his scientific background and interests. Project scientists have to figure out how to achieve the optimum match between the science needs of a mission and its engineering and budget constraints. They also sometimes have to make tough decisions about which experiments and which observations will or will not be done. If members of the science team can't agree on how to carve up available resources (power, time, data volume) for competing measurements, it is the job of the project scientist to step in to arbitrate, or to just plain decide.

"It turns out," Ed reflected, "that's a much more critical role than I had thought ahead of time, and that's because ultimately what science is all about is making discoveries. By deciding to make this observation rather than that one, you're effectively deciding that that group of scientists gets to make a discovery and this group doesn't."

Ed Stone's impeccable record as a careful and thoughtful scientist, his patient and friendly demeanor, and his ability to work fairly with ten project managers and hundreds of *Voyager* scientists and engineers have established him as an effective and respected project scientist, as well as a widely recognized spokesperson for the entire Project during press conferences and media appearances. *Voyager* is run by committee, and consensus is most often the ruling doctrine. But if things were different, I can easily envision Ed Stone as the king of *Voyager*, ruling benevolently over an empire that extends out to the farthest reaches of the known solar system, and beyond.

REMOTE SENSING

Astronomers traditionally study stars or galaxies; geologists typically study rock outcroppings or map oil and mineral deposits; meteorologists study the weather and climate and try to make forecasts—these are relatively traditional and established fields of scientific study. But what do you call a person who studies the science of the planets, moons, asteroids, and comets around us and has to use the theories and methods of some, many, or *all* those fields at once? The term "planetary scientist" is a relatively new one among academic professions, and it's one that the *Voyagers* have helped establish. We're a sort of jack-of-all-trades kind of people, thinking about science questions on scales from the planetwide (like from a mapping camera on an orbiting spacecraft), to the minuscule (like individual little rock piles or sand piles studied by a rover). Some of us are more interested in astrobiology—the study of life in space—than anything else. A common approach among planetary scientists is to use *remote sensing*, very remote sensing, to do our science, because except for those lucky dozen astronauts who've had the privilege of walking on another world, none of the rest of us can actually *set foot on* the places we're studying.

We use technology to experience the place remotely. Everyone actually uses remote sensing of a sort all the time in our daily lives, using our senses to interrogate the world out of our reach to, for example, judge distances and sizes or to identify objects from their shapes or colors or smells. All animals do it one way or another; plants, too. The difference for planetary scientists is the use of robotic sensors: cameras acting as eyes to provide sight, spectrometers or sampling probes acting as organs for smell and taste, arms and scoops and drills providing a sense of touch, radio antennas for "hearing" and "talking." And even a sixth sense comes into play sometimes, one that is familiar to hikers or geologists working out in the field: a sense of place or context enabled by mobility—the ability to roam and climb and explore a place from multiple perspectives, or to leave it entirely and head for new ground. Flyby and orbiter spacecraft, and rovers on the ground, provide these essential remote-sensing capabilities for us, sending back the pictures and sensory data from remote places.

It's easy to think of spacecraft like the *Voyagers* as being alive, imparting to them feelings and other human attributes. They are so far away, and it is so cold and dark. They must be lonely. Some of them, like the Mars rovers *Spirit* and *Opportunity*, are so cute with their long necks and bulging eyes! They must be plucky, intrepid, courageous, and a dozen other grand adjectives of exploration, in order to survive and thrive for so long. They are out there, working tirelessly, making discoveries and braving dangerous environments with no rest, no vacation, and no pay. We've got robots exploring the solar system for us!

Well, as fun (or creepy) as that is to imagine, it misses the point. They are machines, built and launched and *operated remotely* by smart and clever people. Spacecraft like the *Voyagers* are high-tech, to be sure, but not sentient or any more capable than their relatively primitive (by our twenty-first-century standards) software. "Don't anthropomorphize the spacecraft," *Voyager* imaging team member Torrence Johnson recalls Project Manager John Casani saying. "They don't like it."

"The sense of exploration we get with these missions is a very 'human explorer' kind of feeling, even though our senses are on the distant spacecraft," my friend, planetary science colleague, and *Voyager* imaging team member Heidi Hammel says. "I feel like an old-fashioned mountain climber when I am making discoveries, seeing something for the first time, realizing that no human before me has ever seen what I am seeing. It takes your breath away—for just a moment you feel a pause in time as you know you are crossing a boundary into a new realm of knowledge. And then you plunge in, and you are filled with childlike joy and wonder and delight." Like me, Heidi cut her teeth in our business with *Voyager,* and like me, she got hooked on the thrill of exploration and discovery. "And then you get serious again, and start thinking about how it fits into what you already know," she continued, "and your grown-up scientist brain takes over. Those of us who have had that feeling want to keep coming back for more, and we want others to have that feeling too."

A PLANETARY SOCIETY

The robotic exploration of space is, in fact, human exploration. It's just that the humans doing the exploring haven't left this planet. And that's why the story of the *Voyagers* and their travels to the edge of the solar system and beyond is a story of the human drama of deep-space exploration. *Voyager*'s saga is one of discovery and adventure but also of risk and frustration, successes as well as sacrifices, consensus and conflict, the historic mingling with the mundane. Scientists, engineers, managers, technicians, artists, students, and countless other professionals designed the mission and built the spacecraft, guided them on their Grand Tour of the outer solar system, helped take the photos and make the discoveries that now fill our textbooks, and still help us communicate with the spacecraft today on their continuing interstellar voyages. When historians five hundred years from now look back, the accomplishments of this particular group of people will be among the most important remembrances of our time.

But many other people have had an important albeit indirect role as well. As a student, I learned about and joined a new organization that my hero Carl Sagan had helped form in 1980 called The Planetary Society. The Planetary Society is the world's largest public-membership space-advocacy organization, and its beginnings are as tied to *Voyager* as my own. America in the late 1970s was in a state of national crisis: high inflation, high prices (and even rationing) of oil and gas, and federal budget deficits rising to levels not seen in decades. Ronald Reagan was elected president in 1980 partly as a result of a national backlash against President Jimmy

Carter's administration's inability to get the economy back on track. Reagan interpreted his mandate to be to recover the economy by promoting business growth (this is when the term "Reaganomics" was coined) and cutting taxes and federal spending. Specifically, that meant cutting *nondiscretionary* federal spending—the programs not related to defense or Social Security or Medicare and other entitlements. NASA found itself on the chopping block for massive potential budget cuts, initiated by Reagan's chief of the Office of Management and Budget (OMB), David Stockman.

I don't know whether Stockman liked NASA or not, but it didn't matter—even though its fraction of the federal budget was less than 1 percent (as it is today—now less than 0.5 percent, in fact), the space agency was an easy target for budget cutters. "Why should we spend money on searching for little green men," some in Congress have asked (*really*), "when we have so many other pressing needs here at home?"

Why should American taxpayers support NASA? I believe it's because we like satellite launches, space shuttles, moon landings, Mars landings, and cutting-edge materials and computers and communications technology and products . . . even Tang, at least for a while. Most important, however, as is obvious to anyone who spends time around scientists like Ed Stone or Heidi Hammel, or science popularizers like Bill Nye or Neil deGrasse Tyson, there are critically important *intangibles* in this pursuit that feed our souls. Some of the intangibles of supporting NASA—results for which we cannot predict their future influence on our society or our planet— include the inspiration and education of our young people, the gathering of pure knowledge about the worlds around us and our place in the universe, and of course national pride in American leadership in exploring the greatest frontier there is.

The national debate and budget-cutting angst of the early 1980s was happening right on the heels of the spectacular *Voyager* flybys of Jupiter in 1979 and Saturn in 1980. These planets had been visited before, though only briefly, during flybys by the *Pioneer 10* and *Pioneer 11* spacecraft a few years prior to *Voyager*. While spectacular achievements in many other ways, the *Pioneer* images of the giant planets were somewhat fuzzy (not much better than telescopic images from Earth, because of the relatively far flyby distances and crude digital imaging technology that was used), and little new information was obtained about the moons or rings around those worlds.

Voyager was different.

Van Gogh–like tapestries of crazily colored, swirly clouds with vibrant tones of orange, yellow, and red on Jupiter, including the first close-ups of the Great Red Spot, began showing up on TV, on space posters, and in textbooks. The clarity of form and color and the simple elegance of Saturn's rings were revealed for the first time, including photos *looking back* from behind the rings, beyond Saturn, viewing the planet from a perspective impossible to achieve from Earth. And the large moons around Jupiter and Saturn were unveiled as alluring worlds—planets in their own right—one with active volcanoes (Io), another with plates of what appeared to be floating sea ice (Europa), and another with a thick, smoggy atmosphere that may be what the Earth's early atmosphere was like (Titan). It was a grand spectacle.

Voyager imaging team member Carl Sagan knew, from his own experience as a public speaker, educator, and TV host, that there was enormous public support for NASA, but that it was scattered across the country and not organized in any particular way. Something had to be done to combat the looming budget cuts. Sagan, along with Bruce Murray, a planetary scientist and then director of JPL at Caltech in Pasadena, and JPL space mission engineer and manager Louis Friedman, decided to try to organize and focus public support. In 1980 they formed a nonprofit membership organization, a society that any like-minded people could join. Dues would be $15 per year, and they'd send out a bimonthly magazine with the latest space images and other related information. They called it The Planetary Society, and the magazine *The Planetary Report.*

I joined The Planetary Society as a high school student in 1980 (I think I saw the ads for membership in the materials distributed by my rural Rhode Island amateur astronomy club, SkyScrapers). Many members of the club were as excited as I was about the *Cosmos* TV show and thrilled about being part of a nationwide—worldwide— effort to promote space exploration. I let my membership lapse a few times in college when I was short on cash, joined again for good in grad school, and now I'm privileged to be the president of the society's board of directors, a position once held by my mentor, Carl Sagan. Membership in the society skyrocketed to more than 100,000 people shortly after it was formed, partly fueled by Sagan's enormous popularity and influence, but partly also because it provided a way for interested people to stay informed about and connected with the space program in the days before the Internet. Sagan, Murray, and Friedman took this public support to Congress and the presidential administrations over the years to help demonstrate the high level of enthusiasm for NASA and space exploration. Indeed, the society was instrumental in tapping into the success and legacy of *Voyager* to help avert the worst of the early 1980s budget slashing of NASA and to help set the stage for the phenomenal missions of exploration and discovery that have taken place since.

Today, thirty-five years after it was founded, more than a half million people have been members of The Planetary Society, and millions more enjoy the images, activities, articles, blogs, and tweets on our website for free (planetary.org). And once again we find ourselves in austere times, where shortsighted members of our government are again looking to slash and burn federal budgets for nondiscretionary programs like NASA. So once again we are rallying the troops, tapping society members (and nonmembers as well—indeed, anyone else who cares) to write letters and e-mails to Congress to express their support for the space program. Astrophysicist and science popularizer Neil deGrasse Tyson, host of the recent television remake of *Cosmos* and another former president of the society, has called for an increase in NASA's budget *despite* the nation's difficult financial situation, because space exploration represents the very best of what our species does, and because he knows that the value of inspiration is priceless during tough times. The society's new CEO, the Emmy-winning TV host, engineer, and science educator Bill Nye ("The Science Guy"), is reaching out to new members, taking advantage of contemporary venues like social media, to help keep the society vital and effective.

The missions that have come since *Voyager*, such as the Jupiter orbiter *Galileo* and the Saturn orbiter *Cassini*, have revealed those worlds and their rings and moons anew, with more powerful senses, at higher resolution, and over extended periods of time. The time spent by these spacecraft in the Jupiter and Saturn systems has allowed us, for the first time, to see those worlds *in motion*, active and evolving—by nature a difficult task for a flyby mission like *Voyager*. These worlds are dynamic, a fact easy to forget when all you get is a short movie of the approach or a few still snapshots of a place apparently frozen in time. The more extensive time-lapse movies that we now have of these worlds from orbiting spacecraft have brought them to life. We can close our eyes and see the colossal storms raging on Jupiter as they morph into reality. Closer to home, rovers like *Spirit*, *Opportunity*, and *Curiosity* and the orbiters high above them are helping us unlock the secrets of ancient, Earthlike Mars, while orbiters circle and map Mercury, Venus, the moon, asteroids, and comets (as well as our own planet!) to help put the story of our origins together. Space exploration used to be dominated by the United States and the USSR, but now it has expanded into a truly global enterprise with significant contributions from Europe, Canada, Japan, China, India, and others. We are in the midst of a golden age of the exploration of space by people across our planet. About thirty active robotic missions are out there plying the ocean of space on our behalf, poised to make some of the most profound discoveries of all time. These missions let us vicariously see and hear and taste and touch the dirt and wind and ice of other worlds, following in the footsteps of the most grand and far-flung of them all, the *Voyagers*.

VOYAGER AND ME

We first crossed paths professionally, me and *Voyager*, when I was a college student in the 1980s, searching for a way to make a career out of my childhood dreams of astronomy and space. I applied for and—to my amazement—was accepted by both MIT and Caltech to study astronomy (as my friend Bill Nye would joke about his own acceptance by his beloved Cornell, "There must have been a clerical error of some kind"). Against the wishes of family and friends (most of whom had never heard of the place), I chose Caltech, partly because I needed to spread my wings and explore firsthand the things I was hearing about this strange new world called California, and partly because I knew that Caltech was intimately connected to JPL, the epicenter of planetary exploration in the United States.

The smell of the olive trees the first day I walked onto the campus of the California Institute of Technology in the fall of 1983 was the smell of *newness* and *change*. It turned out I had traded an insular, small town and small-state family life for an insular, small dorm and small-campus nerdy life. I had never been so challenged academically (the professors are notoriously merciless there, since many have either invented the field they are teaching and written the textbooks themselves, or they are busily distracted by, and actively working on, the cutting edge of whatever was being taught). I failed Math 1 and so they put me and a small group of other struggling students in a "special" math class called Math 0.9—just slightly less than Math 1.

One day, I saw a small ad posted on one of the bulletin boards that I would frequently peruse on my job search around campus (remember: no Internet!), looking for a student to help analyze some ultraviolet measurements of Jupiter.

That sounds like astronomy, I thought to myself. Why not check it out?

The ad was posted by Mark Allen, a Caltech/JPL research faculty member. He grilled me in his strong New York accent about my background and previous experience during the interview. Despite consistently hearing myself answer "no," "none," or "I don't know what that means" to his questions, I felt good about him. Up to that point, my life's work experience had consisted of picking car parts in my father's junkyard, plotting chemical assay results in a metal refinery lab run by the father of one of my high school friends, and taking out the trash along with other odd jobs in a mall clothing store for large women. Incredibly, Mark offered me the job anyway. Maybe it was the plotting experience (that's mostly what the work he needed turned out to be about). Or maybe it was a clerical error. Whatever the reason, it changed my life.

Working every day in South Mudd, Caltech's building that houses its Division of Geological and Planetary Sciences, was an absolute delight for a young space junkie. There were posters and murals and space paraphernalia all over the place, and the halls and offices were lousy with famous (to me, at least) faculty, staff, grad students, and postdocs who were working on missions like *Voyager* and *Viking*. I would faithfully print and analyze plots for Mark, but while waiting for my plots to print or for my computer job to process (I was low on the priority totem pole), I would wander the halls and daydream about the far-off places the people around me were exploring. It was *much* more fun than classes and homework and exams, and so my grades continued to suffer. I kept my head *just* above the waterline and was lucky not to "flame out" like some of my other fellow Math 0.9 friends—but it was close.

I grew up thinking that astronomers studied everything: stars, galaxies, planets, moons, whatever. It's all in space, right? But it turns out they're compartmentalized, balkanized, self-segregated by distance, and then by energy: solar system, galactic, extragalactic, cosmologic, and then from microwave/infrared (low energy) to

UV and gamma rays (high energy) within each of those realms. At a party with these people, you would not want to confuse a high-energy extragalactic cosmologist with a near-Earth asteroid hunter, believe me! But I also learned that there are *different* kinds of *solar system* researchers out there who are *not astronomers* but who are geologists, or chemists, or physicists, or meteorologists. They happen to study nearby solar system objects (or maybe meteorite samples of them) rather than astronomical objects, and at Caltech at the time, most of them didn't use telescopes but instead used images and other data taken by robotic space missions to do their science. Many of them were designing or flying their own cameras and other instruments in space. I had found my tribe! It turned out that the particular flavor of astronomy, space mission, and hands-on engineering experience that I was looking for had a name: planetary science. And at Caltech, I could actually get a *degree* in planetary science! I switched my major.

The halls of South Mudd are where I first met G. Edward Danielson Jr., or just "Ed," as he liked to be called. A cheerful, big, sometimes shy gentleman, Ed was a member of the Caltech/JPL technical staff who specialized in designing, building, and operating cameras in space, for missions like *Mariners 6, 7,* and *10, Viking,* and the Hubble Space Telescope. He was also a member of *Voyager's* imaging team and spent a lot of time looking at and analyzing the incredible images sent back from Jupiter and Saturn just a few years earlier. I would run into Ed at the printer, where I would often have to sheepishly hand him back his printout—which I was holding and admiring—of some amazing *Voyager* image of Saturn or *Viking* image from Mars that had printed out before my plot. He started getting into the habit of giving them back to me, saying something like "Oh, that's the wrong one" or "Oh, the contrast is wrong in this one" or some such. But I was onto him—he knew how much I treasured each of those printouts, even the underexposed or oversaturated ones. It was a fun little game, and soon I felt comfortable enough to talk with him about what he was doing. That's how Ed turned me on to a new and growing field called *image processing*.

While I was too inexperienced and naïve to know it at the time, I later came to realize that at his core Ed was really a tinkerer, more of an engineer and a *science enabler* than a pure scientist. He cared about helping to make science discoveries from the *Voyager* images, for sure, but he cared more about how the cameras were working, so that he could help make sure that they were taking the best possible photos that could be taken. For a flyby mission, you get only one shot—so you want to make sure the cameras are pointed in the right direction, and you have to make sure that the exposures are at the right level. Point toward empty space or set the exposure time too short: all-black image. Point in the right direction but take too long an exposure: all-white image, or at least lots of uncorrectable saturation. Accidentally point at the sun: fry the camera. The stakes were high, and so guys like Ed who were responsible for getting it right *had to get it right*. I had never encountered that kind of pressure among scientists or engineers before.

As Mark Allen's project was finishing up, it turned out Ed Danielson was looking for some help processing images from the *Voyager 2* flyby of Saturn so that they could understand how the camera was changing over time, as it got older and farther away from the sun, to help the team prepare to *get it right* for the flyby of Uranus the following year. I will never know if Ed really needed a student to do that work or if he made the job up for me to keep me around. I was delighted and jumped at the chance. It was like hitching another ride on the *Voyagers*.

Back in the day, image processing was done on *the* (single) computer used by each faculty group (it was, conveniently, located near *the* printer), and it was a pretty competitive ordeal to get time to run programs and analyze images. As an undergraduate working among a group of active and productive faculty, postdocs, and grad students, I was the last guy in the queue, and so to get time on the image-processing computer, I often had to come in to work during the graveyard shift, catching Ed at either the end of his workday or the beginning of the next. Those nights were ghostly quiet, with just the hum of the nearby computer fans or the distant whine of the janitor's vacuum cleaner to keep me company. As I gazed upon image after image from *Voyager*'s close flyby of Saturn, my thoughts would sometimes wander. I'd imagine myself as a passenger on that ship, imagine the gasps my fellow travelers and I would make as we passed through the flat disk of Saturn's glorious, gossamer rings. And flat indeed! Although the main rings span the width of more than twenty Earths, they are only about thirty feet thick! If Saturn's rings were a DVD, that DVD would only be about ten *atoms* thick, or about 100,000 times thinner than a human hair. The cool thing was that no one knew exactly why they were so thin, but I figured the answer was probably right there among the images I was working with.

I used to tell my friends that I was working at the edge of space. That's because my job was to pore over the *Voyager* images and, literally, find the edges of space—the pixels where the planet ended and space began. Many of *Voyager*'s images, especially the ones taken when the spacecraft was really close to Saturn, have parts of the planet's edge (called the *limb*) or the rings' edges gracefully arcing through the photos. I would identify those parts of the image, and using some special software that Ed and others on the *Voyager* team had devised, I would try to fit a smooth mathematical curve to those edges. The curve was an estimate of how the edge of the planet *should* be curved when viewed from *Voyager* at the time and place that each photo was taken, if *Voyager* were precisely where it was predicted to be and if the camera were behaving exactly as predicted. But the spacecraft was *never* precisely where it was predicted to be, because of the slight push and pull of the gravity of Saturn and its moons, and the cameras *never* behaved exactly as predicted, because of the strange ways that the cold temperatures or the intense radiation from the magnetic fields close to Saturn itself could introduce artifacts into *Voyager*'s old-style vidicon camera system (which, unlike a modern digital image detector, would capture images using a cathode ray tube and electron scanning gun, like in old TV sets). So my lovely curves would never fit perfectly the first time, and I'd have to go back and fine-tune them a little to get a better curve to fit the limb, nudging the inferred position of the spacecraft a little, or tweaking some aspect of the lens distortion.

It was interesting work but it was painstakingly slow. Just to *display* an 800 x 800–pixel *Voyager* image on Ed's screen could sometimes take thirty seconds to a minute (depending on who else was using the computer's processor), as the pixels slowly scrolled down onto the screen like thick paint dripping down a wall. To run the limb-fitting program, I had to submit the program as a "job" to the computer, and that job could be in a queue with other jobs for hours. I had to give my job a priority: low numbers like 1 to 5 were highest priority, and high numbers like 15 to 20 were lowest priority. I could see who else was running jobs on the computer—many professors and students would submit big complex calculation jobs in the 15-to-20 range at the end of the day so they could run overnight. If I was there at night working alone, I'd often give my jobs high priorities, a 3 to 5 maybe, so they'd run quicker (maybe a half hour per limb fit, if I was lucky). A couple of times I messed up, though, leaving my high-priority job caught in some infinite mathematical loop and still accidentally running during normal daytime hours. I'd get some dirty looks in the halls when I returned later in the day. I suspect Ed got chewed out over my screwups more than once.

One day late in the fall semester of 1985, Ed came to check in on me before leaving for the day and he casually mentioned that he was getting excited about *Voyager*'s upcoming flyby of the planet Uranus in January. He told me that, using their detailed calculations, the JPL professional celestial navigators on the team were hatching a plan that would enable a very precise flyby of the planet and its moons. Everyone felt good about being able to use the Uranian gravity to slingshot *Voyager* on to Neptune. My limb-fitting work was only a tiny piece of the planning puzzle, but still it felt good to know that I was making some contribution, however small. Then he asked—and I'll never forget this as long as I live—if I would like him to put my name on a list to get me a special badge that would allow me to witness the encounter from JPL, from *inside the Science Operations Room*. Um, yeah. That would be great, Ed. Thanks! I did cartwheels all the way back to my dorm at sunrise the next morning.

Ed got me that badge. I still have it.

IN THE ROOM

Ed's magic badge got me into the famous Building 264 at JPL, where the *Voyager* scientists would get their first look at these long-awaited images as they streamed in from NASA's giant Deep Space Network radio telescopes around the world. All throughout January we could tell that *Voyager* was getting close to its target. We were all mesmerized as Uranus went from a fuzzy dot to a recognizably round and beautifully blue-green Ping-Pong ball as we approached. Now, Ed had given me the badge, but he hadn't given me any sort of job to do. He wasn't even around in the main science team work areas; instead, he was busy working out various last-minute issues with the camera pointing or exposure times for the flyby sequences, sequestered somewhere with a small band of critical mission planners who needed quiet spaces and uninterrupted time to do their calculations. So I was kind of an interloper, a fly on the wall, with no real reason to be there.

I tried to make myself useful in some way. Images were streaming down day and night from the spacecraft as it got closer to Uranus, and over the days leading up to *Voyager*'s closest approach the room started to fill up with planetary geologists, atmospheric scientists, space physicists, and their students and staff. I glimpsed Carl Sagan a few times but never got a chance to formally meet him, because he was usually surrounded by a gaggle of others. I met some of the grad students who were working with the images for their thesis projects, and they were like kids in a candy store. I met some other planetary scientists on the *Voyager* imaging team, such as Carolyn Porco (studying rings), Larry Soderblom (studying icy satellites), and the late Gene Shoemaker (one of the fathers of modern planetary geology, studying everything and everyone), although no doubt they don't remember meeting me. A few other lucky undergrads (such as Heidi Hammel) and I would hang out in the back of the science rooms and look for people who might need a coffee, or a photocopy made, or someone to run out and grab some sandwiches or meet the pizza delivery guy by the main gate. In fact, my primary contribution to the *Voyager* flyby of Uranus may have been to help keep some of the key team members from dying of starvation. I didn't care—I was *in the room*.

There's an interesting sociology and psychology to be witnessed when groups of people are placed together in stressful situations. Some of the people in that room had been working on this project for more than a decade, preparing intently for each of the precious few moments when the *Voyagers* would fly by their targets. The pressure would build as each person pondered the inescapable fact that there would be only one shot, one chance to get it right. Some people handle stress gracefully, while others don't, and I saw or heard about plenty of examples of both within the inner sanctum of the *Voyager* science and operations areas as the encounter got closer.

The PI of the imaging team, a planetary astronomer from the University of Arizona named Brad Smith, was a force to be reckoned with: sometimes jovial, other times stern, and clearly—like everyone else in the room—worried about it all going well. He kicked us interlopers out several times, for special "team members only" meetings or just because he thought there were too many people in the way. Sometimes he was nice about it; sometimes he wasn't. My former Cornell University colleague and research mentor Joe Veverka, a member of Brad Smith's *Voyager* imaging team, tells a story about how, at an early team meeting, Brad had everyone shake hands and apologize to one another in advance for the nasty things they were probably going to say and do in the heat of the upcoming stressful flybys. Joe would later give the same advice to the imaging team that he led on NASA's Near Earth Asteroid Rendezvous mission, and as a member of that team I can vouch for Joe's (and Brad's) sage advice.

I would see Project Scientist Ed Stone, who seemed to me, from the back of the room, to have that gracious and kingly sort of personality, make little speeches now and then to encourage the team, or intervene and occasionally break up some of the somewhat-too-heated debates that would pop up among the tired and overworked instrument teams. I asked Ed how he kept his cool during those stressful, exciting days.

"I don't know!" he said. "It was just such a great thing. There was so much hard work; there were discoveries every day. It was just incredible. And wonderful."

I had to conclude that Ed must have just had the right genetic disposition to lead a group of talented and highly motivated people through such stressful experiences. My friend and colleague Ann Harch, who was a *Voyager* sequencer and the science operations coordinator during the Uranus and Neptune flybys, says that she and others viewed Ed Stone as an incredibly fair and approachable leader. "He did an amazing job of making sure that *all* of the science investigations got their important science into the plan," she recalls.

The Uranus flyby itself, on January 24, 1986, went stunningly well (much better than the Super Bowl a few days later, in which, sadly, my Patriots got crushed by the Bears 46–10). Uranus itself was fairly bland, showing no evidence of the stunning clouds and storms like Jupiter and Saturn. It was the Uranian moons that stole the show—tiny, icy worlds with staggeringly high cliffs and cracks and super-dark plains interspersed with bright (icy?) cratered debris. Uranus is tilted on its side (rolling around the solar system instead of spinning), and so its dark rings and five large moons make for a sort of bull's-eye dartboard pattern that *Voyager* was aiming for. The spacecraft passed through the bull's-eye close to the tiny, jumbled-up moon Miranda just before making its closest approach to Uranus. Images streamed in nonstop. A few of them would later make their way into newspapers or onto the network news, but those of us there at JPL were the fortunate ones who got to experience, live, our first encounter with these exotic worlds.

And then, just like that, Uranus was in the rearview mirror. We watched the blue-green world that we'd been seeing head-on for weeks wane into a thin, ghostly crescent as we passed behind it. The planet's rings were revealed in all their glory by looking back at them, into the sun, making their tiny particles light up for the cameras like a wet or frosty windshield lights up when you're driving into the sunlight. Many team members started to pack up to head back to their homes, with data tapes and stacks of photos in hand and dreams of discoveries to make and scientific papers to write.

Ed Stone and other *Voyager* team leaders began to prepare for a NASA press conference in which they would share the "greatest hits" from the Uranus flyby—stunning images and other measurements of the planet, moons, and rings, as well as initial results about the planet's magnetic field and its interactions with the solar wind. Ed Stone thought it would be a wonderful celebration of one of the greatest achievements to date of the space age—the first encounter with a new world.

But the celebratory mood and all that press-conference planning stopped suddenly on the morning of January 28, when the Space Shuttle *Challenger* exploded just seventy-three seconds after launch, killing all seven crew members. I remember it vividly, watching on TV from my college dorm room. I never missed a shuttle launch on TV, and I was even lucky enough once to hop in a car with a bunch of college friends at the last minute and catch a shuttle landing at Edwards Air Force Base in the Southern California desert. Seeing *Challenger* explode live on TV was jarring, not just for me but for my *Voyager* colleagues, everyone else involved in the space program, and the nation as a whole. Management and design flaws in the shuttle system were uncovered, and it forced major debate and soul-searching about the importance of human space exploration for America's future. The media became instantly consumed by the *Challenger* disaster, fleeing Pasadena in droves and leaving the rest of the *Voyager* Uranus story untold. Ed Stone and the rest of the project leaders knew they had to postpone their press conference. The "greatest hits" made it out there, eventually, but they were released without as much fanfare. The remaining somber team members slowly drifted off and headed back home.

A few weeks later, back on campus, I finally had a chance to see Ed Danielson again for more than a fleeting hello. He had been working hard as an imaging team liaison to the JPL navigation team to figure out how much *Voyager*'s trajectory had been bent (ever so slightly, but measurably) by the gravity of Uranus and its moons. This would enable the navigation team to estimate the mass of the moons, which, when combined with estimates of the moons' volume from the images themselves, would let the team estimate their densities. The moons turned out to have very low densities, close to that of water ice.

Perhaps not surprising, given their location in the cold outer reaches of the solar system, but still, Ed wanted to get the numbers right. I tried to express the depth of my gratitude for getting me that badge.

"It was nothing," he said.

After graduation, Ed and I stayed in contact, catching up at various conferences or during my occasional trips back to Caltech. He took a leading role in the development of the first high-resolution planetary camera, the Mars Orbiter Camera (MOC) on the *Mars Observer* mission. Unfortunately, that spacecraft, and Ed's darling camera, blew up just three days before getting to Mars. Undaunted ("It was nothing"), Ed and his MOC teammates from Malin Space Science Systems in San Diego built another one a few years later for the *Mars Global Surveyor* mission, and *that* MOC got to Mars safely and went on to discover gullies and deltas and massive sedimentary layers—photos of which would forever change our perception of *that* world as well.

Ed retired in 2004 and, after battling complications of a stroke, passed away in late 2005. I still feel his influence on me every day. Using the skills Ed taught me as we pored over those early *Voyager* Saturn images in his workroom, I was later able to work on a project of my own, mapping the geology of Miranda and the other moons of Uranus. I also developed a sense of the important role that hardworking scientists like Ed Danielson and Ed Stone can play behind the scenes in the enterprise of space exploration. The grunt work of science—planning the observations, calibrating the images, processing the data, making the mosaics, training the newbies, balancing the budgets, and so on—is critical for the team to *get it right*. Missions like *Voyager* succeed because of people like them. This world needs its tinkerers as much as it needs its theoreticians.

2

Gravity Assist

As a kid launching model rockets in my backyard, spending hours carefully gluing parts together, applying stickers, painting the fuselage, packing the parachute, and installing the engine, I would wonder, *Did I balance it right so it will launch and fly straight?* Most times it did, but sometimes it never got off the pad or launched sideways, causing me and my sisters to run for safety. *How far will it fly?* Sometimes completely out of sight, never to be found, maybe swallowed up by the trees ... *Can I figure out how to mount a small film camera to the nose cone?* I never did, too heavy ... Repairs or rebuilds had to be done, then another long lead-up to the launch, and then another suspenseful countdown as family and friends stood by—often watching from indoors to stay safe.

It turns out that many of the problems that need to be solved to launch model rockets are the same kinds of problems that engineers and scientists involved in NASA space missions have been working to figure out since the 1960s. How do you design the craft to withstand the stresses of launch and the harsh cold vacuum of deep space? How do you figure out how to communicate with it and control it once it's out there? How do you get pictures and other measurements sent back from it? How do you design its mission? Early on, there was another question that arose once we started thinking about space travel: can we use a planet's gravity to speed up and turn a corner? That could enable our rockets, our spacecraft, to go even farther. . . .

Even though it's a cliché, lots of times a space mission really *does* start with scribbles on the back of a cocktail napkin. Or with conversations among colleagues over beers after a day at a professional conference. Or sometimes the idea comes in a dream, or in a flash of realization akin to making a discovery. Indeed, the *Voyager* mission appears to have begun in such a flash of inspiration.

In the mid-1960s, Gary Flandro was a graduate student in aeronautics at Caltech studying instabilities in rocket combustion and working part-time at JPL helping to study the aerodynamics and trajectories of missiles. His supervisor was one of the key members of the JPL Mission Analysis Group that was working to devise the upcoming *Mariner 10* Venus–Mercury gravity-assist flyby mission, and he suggested to Gary that he help explore the possibility of similar gravity-assist trajectories being used for outer solar system missions. It was an area that almost no one else was looking into yet at JPL, as the lab was focused almost exclusively at the time on lunar, Mars, and Venus mission work.

Gary was also a fan of rocketry and of the history of the

academic field known as celestial mechanics—the calculation and prediction of the orbits of planets, moons, asteroids, comets, and eventually spacecraft. In an interview published by *Voyager* mission chronicler and University of Hawaii sociologist David Swift, he pointed out that "the basic ideas behind gravity assist were known as far back as the 1800s." These ideas were partially based on analysis by early celestial-mechanics pioneers, such as Urbain Le Verrier from France, of deviations of the orbits of comets passing by Jupiter. Le Verrier would later go on to use the same calculations to deduce in the 1820s that the planet Uranus had performed a distant gravity-assist flyby of a massive but as-yet-unseen object, which altered its orbit slightly. With Le Verrier's help, that massive mystery object was eventually revealed to be the planet Neptune. It was on the shoulders of such giants that Gary Flandro began to search for similar ways to use gravity boosts to speed up the decades-long interplanetary travel times that would be required for direct-from-Earth outer solar system robotic missions.

One of Gary's goals was to find out if gravity assists could be used to get to Saturn, Uranus, Neptune, and/or Pluto *quicker* than direct trajectories (essentially by using a close flyby of Jupiter as a slingshot), while still allowing a spacecraft to carry a significant amount of mass for propellant, power/communications/thermal systems, and science instruments. His flash of insight, in the spring of 1965, appears to have been to check if the alignments of the outer planets in the near future could, perhaps, allow not just one slingshot, but multiple slingshots that might permit a spacecraft to be fast-tracked to more than one outer planet after swinging by Jupiter. His investigation quickly showed that there would be a rare alignment of all four giant planets, plus Pluto, on the same side of the

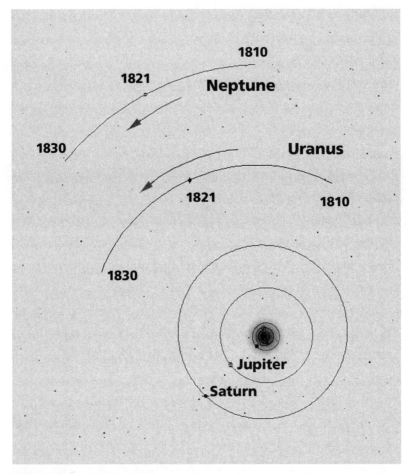

The Uranus Flyby of Neptune in 1821. Schematic diagram of the positions of Uranus and Neptune between 1810 and 1830, showing how Uranus made a distant flyby of the then unknown Neptune in 1821. The resulting tweak of the orbit of Uranus allowed mathematicians to predict the existence of Neptune, and then for Johann Galle to discover Neptune in 1846. (*Jim Bell; SkyGazer 4 [Carina Software]*)

solar system in the 1980s. "So, why not look for a *single trajectory* that would pass each planet with the shortest possible trip time between?" he thought. Indeed, his calculations showed that it would be possible for a single spacecraft to visit Jupiter, Saturn, Uranus, and Neptune, or just Jupiter, Saturn, and Pluto, if it were launched about a decade hence, in the mid-1970s.

The potential efficiency of the gravity-assist process, which had been worked out in detail by others at JPL and elsewhere well before Gary Flandro began his work on the problem, nonetheless appeared in full bloom in his calculations: *nearly twenty years* could be shaved off direct-flight times to Neptune or Pluto by using well-timed gravity assists. Gary could "distinctly remember the feeling of awe as it first dawned on me that this mission was available at just the right time, leaving about ten years to market the mission concept and to design and build a spacecraft. Yes! Here is the way to do it! The next opportunity would not appear until about 175 years later!"

He is still excited about that profound Eureka moment from fifty years ago when he first imagined what JPL's chief scientist at the time, Homer Joe Stewart, would dub "The Grand Tour." "Wow—bang! There it was, right there! This is wonderful!" He describes the feeling of "flying this whole thing in my mind," imagining, for example, a spacecraft traveling between Saturn and its innermost ring in order to get the maximum possible gravity assist. "When it actually happened later, it felt like, 'I've already been there!'" Back in the summer of 1965, Gary excitedly described his results to his boss, who encouraged a more detailed study of the opportunity.

It was a spectacular cosmic coincidence that the technology needed in order to perform such a Grand Tour mission happened to have developed to a sufficient degree on the small planet from which the mission would be launched at precisely the time when the planets would line up just right to make it possible. Through the halls of JPL, Gary's trajectory for a Jupiter, Saturn, Uranus, and Neptune flyby, augmented and enhanced by the work of others, became the "Grand Tour" trajectory. By July of 1965, while JPL's *Mariner 4* probe was making the first robotic flyby of Mars (and while I was busy being born), Flandro had worked out the details of the Grand Tour, including the best times to launch the spacecraft from Earth (the fall of 1977 or 1978).

Even though his multiplanet trajectory calculation work at JPL was far earlier than his PhD dissertation work at Caltech, he still had the good sense to write up his findings and get them published in an academic journal called *Acta Astronautica* in 1966. However, the response to Gary's work was rather negative. "Many openly scoffed at the idea," he said. At the time, JPL was being pushed to the limits to successfully operate missions to the moon and nearby planets that would last a mere several days to several years. The Grand Tour would call for a spacecraft that could operate for maybe a decade or more. Such longevity was unheard-of at the time and difficult for many to imagine. But Gary was now on a quest.

When I was a student at Caltech, one of my friends was a fellow student named Katie Swift, who was also studying astronomy and planetary science. We kept in touch after graduation, partly because she went back to her home state of Hawaii, where I went to graduate school. I got to know her father, David Swift, whose book about the many members of the *Voyager* team included a profile of Gary Flandro.

Despite the fact that the actual trajectories of the *Voyagers'* Grand Tour were selected based partly on Gary's 1965 research, he received little credit for his role at the time. He was still a graduate student, without any real authority or power. His contribution might have gone nowhere—certainly not into interstellar space—and even if it had, his role could well have been completely forgotten. Nevertheless, Gary is pragmatic about the part that he played in making the *Voyagers* happen and, still, rightfully, considers himself a part of the mission. "Many myths have arisen about the origins of the *Voyager* mission," he told David Swift in his interview. "Since I was pretty low on the totem pole . . . no one felt much need to mention my connection with the discovery of the mission. I accepted those misconceptions, but it was sometimes difficult. As a rather naïve young fellow, I could not conceive of the possibility that I would not in time be acknowledged in some fashion for the work I had done. I thought this would happen automatically, since I had properly documented my work and presented it to many people both at the Caltech campus, at JPL, and in various technical meetings." Despite what would be justifiable sour grapes over his lack of recognition, Gary was still gracious about the project overall: "Those at JPL who brought everything together certainly deserve major credit for the magnificent job that they did." It wasn't until 1998 that NASA finally did recognize his contributions, awarding him their Exceptional Achievement Medal. When I first heard this story as a graduate student myself, one line of Gary's jumped out at me: "You have to learn not to be discouraged by experts."

When it came to trying to actually get the Grand Tour mission off the ground, there would be plenty of discouragement to go around. In 1969 and 1970, JPL proposed an aggressive series of missions that would accomplish the Grand Tour. Four spacecraft would be launched: two in 1977 to fly by Jupiter, Saturn, and Pluto; then two in 1979 to fly by Jupiter, Uranus, and Neptune. The spacecraft would be based on JPL's successful *Mariner* series (which by 1969 had successfully flown by Venus and Mars, and which would soon fly by Mercury). It would include probes launched into the atmospheres of the giant planets, and multiple launches per opportunity would help reduce one of the engineers' biggest concerns and risks: keeping the spacecraft alive and functioning for more than a decade. This set of Grand Tour missions was not only ambitious, it was also expensive, with an estimated cost in 1970 of more than $900 million, equivalent to almost $5.5 *billion* today.

Despite the compelling proposal to conduct an exciting and historic mission that would take advantage of the unique Grand Tour opportunity, both NASA and the administration of President Richard Nixon balked at the steep price tag and did not approve the concept. Part of the reason was due to overall NASA budget cuts and belt-tightening as the *Apollo* moon program was winding down (Nixon canceled the planned *Apollo 18, 19,* and *20* missions in 1970), and part of the reason was that NASA was being directed to start ramping up more of their shrinking funding for a new human exploration vehicle called the Space Shuttle.

However, as JPL Grand Tour mission manager Harris "Bud" Schurmeier had recalled in a recent interview on the topic, the door was left a little open. "They told us, 'If you guys can come up with something less grandiose, we'll consider it.' So we went home and quickly put together what we called *Mariner Jupiter Saturn '77* (MJS-77)." Bud and his colleagues at JPL scrapped two spacecraft, designed the remaining two so they would primarily use technology already developed for the *Mariner* series, took off the atmospheric entry probes, and scaled the mission back to just flybys of Jupiter and Saturn. The price tag dropped to about $250 million (about $1.5 billion today), and in 1972 the scaled-back proposal was accepted by NASA and the Nixon administration.

As crazy as it seems in hindsight, NASA managers officially

passed up on the Grand Tour opportunity as Gary Flandro and others had originally conceived it. The *Voyagers* were missions to Jupiter and Saturn, and scientists and mission managers could only hope that at least one of them could continue on to Uranus and maybe Neptune. Ed Stone recalls that "the idea of getting to Uranus and Neptune was being pursued, but quietly, partly because nobody wanted the mission to be considered a failure if we didn't survive past Saturn!" If the Grand Tour was to be resurrected, it would have to be put together in pieces added later—after success at Jupiter, Saturn, and Titan was in hand. Building spacecraft that could fly farther and last longer than any ever had was now the challenge.

CONSTRUCTING THE CRAFT

Once funding for *MJS-77* was approved by Congress and the upper administration of NASA, they handed the job of actually making it happen to the engineers, scientists, and mission managers at JPL. Outwardly, JPL has the look and feel of a college campus, a mishmash of buildings from high-rises to trailers nestled into the (sometimes smoggy) foothills of the San Gabriel Mountains in the small city of La Cañada Flintridge. Tame wild deer roam around under the pine trees and walkways, and the sound of horses and riding instructors can often be heard from the riding stables nearby. JPL began as a US Army–funded rocketry lab on the Caltech campus in the 1930s; as the tests and launches got more ambitious, the facility was eventually transferred to an off-campus location near a more spacious arroyo about seven miles away in what was then northern Pasadena. In the late 1950s, JPL became affiliated with a new federal agency called NASA. Not officially one of the ten "Centers" run by NASA across the country, JPL is instead one of a few hybrid university-government entities called a Federally Funded Research and Development Center, administered by Caltech. JPL employees are actually Caltech employees, not government civil servants, though they all have NASA badges and work with many of the same privileges and restrictions of those on the civil servant pay scale. It is a strange twist of federal bureaucracy that can sometimes create awkward situations, such as when the government closes down over budget disputes in Congress. Some JPLers are sent home; others are deemed "essential employees" who need to stay on the job to protect the taxpayers' investments and keep government spacecraft or facilities running.

JPL has been the epicenter of the American robotic space program since its beginning. Early successes included the *Ranger* and *Surveyor* missions to the moon, the *Mariner 2* flyby of Venus in 1962 (the first spacecraft flyby of another planet), the *Mariner 4* flyby of Mars in 1965, the *Mariner 5* flyby of Venus in 1967, the *Mariner 6* and *Mariner 7* flybys of Mars in 1969, the *Mariner 9* Mars orbiter in 1971 (the first spacecraft to orbit Mars), and the *Mariner 10* flybys of Venus and Mercury in 1974–1975. While JPL had proven itself more than capable in conducting these previous missions, *MJS-77* would be the most complex and advanced planetary mission that the lab had ever attempted.

At its core, every spacecraft, whether on a flyby, orbiting, landing, or roving mission, is built around a basic chassis called a *bus*. The basic starting design of the *Voyagers* was directly inherited from that of the *Mariners*. *Voyager*'s bus is a ten-sided (decahedral) ring-like aluminum structure a little over one foot tall and six feet wide with ten compartments that house most of the spacecraft's electronics and computers. Louvers on some of the ten faces of the bus open and close automatically to help keep the temperature inside relatively constant. In the center of the ring, a pressurized tank is loaded up with 220 pounds of hydrazine (N_2H_4), a common low-thrust propellant used in spacecraft thrusters. The bus design is not exactly the same among all of the *Mariner* spacecraft—some are hexagonal and some are octahedral, but all serve the same basic function of housing and thermally controlling the spacecraft's main electronics components. Similar bus designs were used for *Pioneer*, *Magellan*, *Galileo*, *Cassini*, and even the Hubble Space Telescope. Once engineers find a design that works in the space business, they tend to stick with it.

Building a spacecraft and outfitting it with modern instruments is as complex an undertaking as building an Egyptian pyramid or a Gothic cathedral—and in the 1970s we didn't yet have robots to help fabricate these wonders of human technology. Thousands of people with a huge range of expertise were required in many diverse areas, including mechanical, thermal, electrical, systems, and software engineering; materials science; physics; planetary and space science; fiscal and human resource management; and even in basic hands-on skills such as welding, soldering, sewing, wire-wrapping, and using machine-shop hand tools. People were involved at JPL as well as at other NASA Centers, such as the Cape Canaveral launch facility; at subcontractors and vendors across the country who were providing parts and services and expertise to the core JPL team; and at universities around the world whose faculty, staff, and students were building instruments and preparing to conduct their scientific investigations.

Spacecraft Systems, Ground Data Systems, and Mission Operations Systems subteams were designed to focus specifically on issues like power, thermal control, communications, propulsion, navigation, software, mission operations, instruments, and science. But key managers and bigger-picture thinkers had to be embedded within each subteam to help them work well with other teams. It was known that missions that fail do so usually because the different subteams didn't understand one another's functions or how they must work together. Conversely, successful missions always have outstanding, often paranoid, systems engineers and managers as liaisons.

At the heart of the *Voyagers*, protected inside the main bus from the cold and radiation of space, are three computer systems that control the spacecraft and its instruments. These include (1) the primary computer, known as the Command Control Subsystem; (2) the Attitude and Articulation Control Subsystem (which despite the name is not a mind-control device—it handles propulsion and spacecraft and instrument orientation); and (3) the Flight Data System, whose most important function is to transmit data from the instruments back to Earth. *Voyager*'s computer can process 80,000 instructions per second—cutting-edge space technology in the mid-1970s, but about a million times slower than the laptop that I am using to write this text.

"In everyone's pocket right now is a computer far more powerful than the one we flew on *Voyager*," notes imaging team member and JPL scientist Rich Terrile, "and I don't mean your cell phone—I mean the key fob that unlocks your car."

Still, by splitting the computational functionality into three computer systems, each of which was *reprogrammable* (a feature that was relatively new) and had its own fully redundant backup system, *Voyager* engineers had much more software and operational flexibility than any previous spacecraft. That flexibility would be especially useful when *Voyager 2*'s mission changed dramatically after the Saturn flyby.

The *Voyager* Flight Data System had two ways to get the data back to us humans. *Voyager* could be commanded to radio the data back in real time, essentially as soon as it was captured, and much of *Voyager*'s interplanetary cruise "fields and particles" (nonimaging) data are transmitted this way. Because there were times that the spacecraft were blocked from communication with the Earth, however (as when passing behind a planet or moon), real-time transmissions weren't always possible. So the second way to transmit data relies on using *Voyager*'s onboard 8-track tape player-recorder to quickly record about 100 images and other data from the instruments, and then to play them back later, when the spacecraft was back in real-time contact with the Earth and not busy doing other things. Being able to record and then play back data later made operations and communications more efficient, but it came at the expense of relying on yet another moving-parts system that had to work well for more than a decade in deep space.

It can be surprising just how well what now seems like old technology really worked. I had an 8-track tape player in my car for a time in the '80s, and I remember having a heck of a time keeping that machine from eating my tapes. Sure, all I had were used yard-sale Bee Gees and Barry Manilow tapes, but it was still a hassle to deal with untangling them. Similarly, though not as catastrophically, the tape recorders on both *Voyagers* experienced a number of glitches over the years. For example, when the tape got to the end, it would rewind and start again at the beginning (like an old VCR tape). But this stop, fast rewind, then restart cycle made the entire free-floating spacecraft jolt and wiggle. Even though the forces involved were tiny, the resulting jitter messed up images and other data being taken by the super-sensitive science instruments. As with all the other glitches and surprises, the engineers had to learn to work around these idiosyncrasies.

One of the biggest sources of anxiety that engineers had surrounding the mission, going back to Gary Flandro's original Grand Tour design days, was the need to keep the spacecraft functioning optimally for such long and unprecedented periods of time in the very cold outer solar system. And to top it off, they would have to withstand brief excursions through such dangerous environments as the harsh radiation created by Jupiter's magnetic field, or the probably very dusty and icy environment within the plane of Saturn's rings. A variety of strategies were adopted to mitigate these risks. One was to use shielding made of heavy metals like tantalum as well as radiation-hardened parts to protect the spacecraft from high-energy cosmic ray particles (high-speed protons and other atomic nuclei) in the magnetic fields of the giant planets.

Another strategy was brute-force redundancy of such critical systems as computers, tape recorders, radio transmitters, and receivers. Indeed, shortly after launch, *Voyager 2*'s primary radio receiver failed. The computer automatically switched to the backup receiver, but it, too, began to fail. The *Voyager* team was able to figure out how to avoid failure of the backup, and over time was able to figure out how to communicate with the spacecraft despite its having only a partially working radio and no backup receiver. To assuage their fears of losing the receiver entirely, though, the mission planners had uploaded a small backup mission sequence to *Voyager* that "would hibernate, inside the central computer," says JPL Mission

Planning Office Manager Charley Kohlhase, "such that if we ever lost the other receiver, and could not command any more, at least it had one sequence to carry out: it would get the approach pictures and data for the *next* planet."

Charley was the JPL orbital dynamics engineer and mission architect who winnowed thousands of possible flight paths down to the best handful and who led the design of the critical flyby maneuvers that the spacecraft would use to get amazing images of each planet's atmosphere, moons, and rings, and then slingshot on to the next destination. The flexibility that he helped build into the *Voyager* systems was an important part of overall risk mitigation.

Round-trip travel times for radio signals are many hours between Earth and the outer solar system, and to address this reality *Voyager* utilized a critical and relatively new risk-mitigation strategy called *autonomous fault protection*. *Voyager*'s programmers knew that the spacecraft would be too far away for real-time communication and diagnosis of problems, so they had to devise ways for the spacecraft to recognize problems on their own and to protect themselves from further damage or harm. The engineers who design software fault-protection routines are the kind of paranoid (in a good way) people you'd want with you while you are preparing for a camping trip. Did you pack the tent? How about rain gear? What if the tent starts to leak? And then it freezes? And then the wind picks up? And you run out of water? And there's a bear—no, two! Problem after imagined problem has to be anticipated and a solution thought through. Practitioners of this art talk about exploring every possible branch of the "fault tree"—every conceivable, even unlikely, bad thing that could happen—and having a solution for each situation that saves the day. Having a backup system is one risk-reducing step, but figuring out how to have it switch on by itself when needed was another. *Voyager 2*'s automatic switch to its backup receiver was one example of fault protection in action.

A variety of additional subsystems and instruments were attached to *Voyager*'s bus. These included seven sets of "booms," or appendages of various lengths, extending away from the bus. The longest is a boom made of fiberglass, known as the magnetometer boom; at forty-three feet in length it keeps the magnetic sensor at its tip as far away as possible from magnetic "contamination" by the spacecraft's other metallic and electronic components. The next longest appendages are a pair of thirty-three-foot-long antennas used by the plasma wave and radio astronomy experiments, extending down and away from the rest of the spacecraft. Opposite the magnetometer boom, an eight-foot-long "science boom" holds the Plasma Wave, Cosmic Ray, and Low-Energy Charged Particle instruments along its length, and a steerable scan platform on its end that carries the imaging and spectroscopy remote sensing instruments. By turning the scan platform in different directions, the *Voyager* team could point the cameras and other instruments at targets of interest without having to slew the whole spacecraft. This was a substantial time-saving feature, but it introduced the risks of yet another set of moving parts that would have to continue to work well over more than a decade in deep space. Indeed, a problem with the scan platform on *Voyager 2* would cause some tense and dramatic moments for the team during the spacecraft's passage through the ring plane of Saturn.

The shortest boom holds the spacecraft's radioisotope thermoelectric generators (RTGs), small nuclear reactors that convert energy from the heat given off by the radioactive decay of a few dozen golf-ball-sized spheres of plutonium-238 into electricity to run the spacecraft and instruments. Mounted on top of the bus is a parabolic radio telescope twelve feet in diameter called the high-gain antenna, used for communicating with Earth. And finally, dangling from the bottom of the bus are some triangular struts that look like odd, spindly legs because they're not attached to anything. During launch, however, those struts were attached to an upper-stage propulsion module that helped the *Voyagers* reach their final departure velocity and were then jettisoned.

Before any spacecraft is sent into space, it has to be put through a bunch of tests that simulate the conditions and environments that it will face. These include vibration tables, where the individual instruments and the spacecraft as a whole are violently shaken in the same ways that they will be during launch—and, for good measure, they are shaken *much more* than they will be during launch. For the engineers involved, seeing their creations treated like this can be a painful experience.

The *Voyagers* were assembled from about 65,000 separate parts in JPL's Building 179—the famous "High Bay" Spacecraft Assembly Facility where spacecraft like the *Rangers, Mariners, Vikings, Galileo, Cassini*, and the Mars rovers *Pathfinder, Spirit, Opportunity*, and *Curiosity* were also brought into the world. The High Bay is a Class 10,000 clean room (less than 10,000 particles of 0.5 micron or larger per cubic foot of air), making it a great place to work if you have allergies. Workers in the High Bay have to wear protective clothing (known affectionately as bunny suits) to keep bacteria and other particles (human beings generate millions of skin, hair, dirt, and dust particles every minute) from contaminating the spacecraft.

I've spent time in Building 179, mostly up in the visitors' gallery, but occasionally, luckily, inside the High Bay itself, and to me the place is the closest thing to a modern Gothic cathedral's inner sanctum that I can imagine. Deep inside the High Bay, almost holy relics of our modern civilization are being carefully tended by illuminati who have gone through years of study and training for the privilege of being in that room. They wear ritualistic garb to ensure maximum purity, follow elaborate, carefully prescribed procedures, and when their novitiate work is done, the Chosen One (the spacecraft) emerges from the cathedral and is lofted to the heavens. In that building, and through those doors, pieces of this planet have been worked into structures and systems that are now parts of other planets. Or, in the case of the *Voyagers*, which passed through there too, they are now permanent wanderers among the stars. Building 179 is a factory for Earth's cosmic artifacts, for the things that we cast off this world that will, ultimately, represent us and our time to our progeny, and perhaps even to other beings that we cannot yet begin to imagine. It's no wonder I gravitate to the place whenever I'm at JPL.

Many other supporting facilities are also needed to design, build, and operate spacecraft and missions like *Voyager*. For example, a critical supporting facility for *Voyager* is the Mauna Kea Observatories, built high atop an extinct volcanic peak on the Big Island of Hawaii. Two large telescopes in particular, NASA's Infrared Telescope Facility (IRTF) (with its 120-inch diameter mirror) and the University of Hawaii's 88-inch diameter telescope ("the 88"), were used extensively to provide advance information about the giant planets and their moons in order to optimize *Voyager*'s trajectory and return of scientific data. At nearly 14,000 feet elevation out in the middle of the Pacific Ocean, telescopes there are above much of the warmth and haze and water vapor of our atmosphere and can thus often obtain crisp images of cloud belts and storm zones on Jupiter, Saturn, Uranus, and Neptune or other detailed information on the chemistry and composition of those worlds and their moons and rings.

I did some of my graduate research up there at the IRTF and the 88 and can attest to the harsh conditions of extreme cold, high winds, and low oxygen levels often faced by observers at the summit (often, I would be the only guy on the flight to Hawaii with snow boots and a heavy parka—oh, the strange looks I would get!). Many of the scientists who worked on the *Voyager* camera team got their start as planetary astronomers, obtaining much of the advance information that they needed to plan the giant planet flybys the only way they could—by telescope.

Voyager imaging team member Rich Terrile got his start as a graduate student at Caltech observing "hot spots" on Jupiter with the giant 200-inch diameter Hale Telescope at Mt. Palomar in Southern California. His follow-on work at the IRTF in Hawaii was interesting and relevant to the *Voyager* infrared spectrometer team, which wanted to be able to target some of these "windows" into the deeper Jupiter atmosphere during *Voyager*'s flybys in 1979. Even though Rich was directly helping out the infrared spectrometer team, secretly he was much more interested in being involved with the imaging team instead. "It was much more in line with what I was doing and what I was interested in at the time," he recalls. A chance encounter with imaging team lead Brad Smith on an airline flight led them to strike up a conversation, and then a long-term friendship, that got Rich an invitation to be a full-fledged member of the imaging team from Saturn onward. He and Brad used the IRTF and other telescopes around the world to make observations in support of *Voyager*'s giant-planet flybys, and they have continued their long-term collaboration beyond *Voyager*. For example, they collaborated on a telescopic observation campaign that led to the discovery of the first "circumstellar disk" of dust and gas—a nascent solar system in the making—around the nearby star Beta Pictoris.

Heidi Hammel was a graduate student at the University of Hawaii in the mid-1980s and was doing her thesis research using the 88, collecting color filter images of the atmospheres of the giant planets. With the impending *Voyager 2* flybys of Uranus and Neptune, she was focusing on telescopic observations of those two worlds, and especially on Neptune, which showed cloud features that could be monitored from Earth.

"With my Neptune work, we were trying to establish what the winds of the planet were like," she told me. "Back in those days, we didn't know what the wind speeds were, or even what the exact rotation rate of the planet was. But my thesis advisor, Dale Cruikshank, and I knew that the *Voyager* team would need this kind of information for planning the imaging sequences."

Heidi's results were different from previous studies, and she recalls Dale having her give a "dark, backroom" presentation about her results to *Voyager* imaging team leader Brad Smith and Rich Terrile. She made her case for them to use her results, rather than those from Smith and Terrile's own observations, to plan what images *Voyager* should take.

"I laid out my data on the table, I explained what I did, and I showed how their rotation period just didn't fit the data," she remembers. "I was still a graduate student, and here I was—petrified—

pitching this to the head of the *Voyager* imaging team! When I was done, they just kind of looked at it, and they looked at me, and they said, 'Well, looks like you're right.'"

Smith was impressed enough, apparently, that after Heidi finished her PhD research, he invited her to be a member of the *Voyager* imaging team at JPL.

LEAVING EARTH

Just before they were launched in 1977, what would have been *Mariner 11* and *Mariner 12* if the earlier naming series had continued, the spacecraft were officially renamed *Voyager 1* and *Voyager 2*, partly in recognition of their radically different outer-solar-system missions, and partly because between 1972 and 1977 the spacecraft design had significantly changed from the original *Mariner* configuration.

Once the spacecraft is built and has been tested and proven to be ready for the harsh environment of space, and once its mission and trajectory are defined, someone then has to figure out how to strap it onto a rocket and launch it off the planet. Whenever I'd tried as a kid to launch a small automated film camera "spaceship" on the top of my model rockets, the extra weight proved too much for the engines to lift, and they would either fizzle out or tip over and skip across the lawn (even my monster "Saturn V" model, with five Estes "D" engines, couldn't get more than two feet off the ground!). The lesson I learned, that I now know professional rocket designers have to live by, is that for any given kind of rocket, there is a limit to how much mass can be lifted and accelerated to the required speed, and that mass limit can be only a very small fraction of the rocket's total mass.

In the case of the *Voyagers*, each spacecraft weighed in at around 1,600 pounds (with about 15 percent of that making up the science instruments) and had to be accelerated to more than 25,000 miles per hour to escape Earth's gravity and head toward its first encounter, with Jupiter. The spacecraft were launched in late summer 1977 on the Martin Marietta (now Lockheed Martin) Corporation's Titan III-Centaur rockets, the same kind of rockets that had launched the twin *Viking* orbiters and landers to Mars almost exactly two years earlier. The powerful Centaur upper-stage rocket, a variant of the Atlas Intercontinental Ballistic Missile design of the early 1960s (beating swords into plowshares!), was mounted atop the Titan III rocket and would give the *Voyagers* the big push needed to get them on their interplanetary trajectories.

By a strange quirk of celestial mechanics, *Voyager 2* was launched on August 20, 1977, almost three weeks before *Voyager 1*, which launched on September 5. *Voyager 1* was initially targeted for a Jupiter flyby, followed by a Saturn flyby that included a very close pass by Saturn's large moon Titan, whereas *Voyager 2's* trajectory did not include a Titan flyby. It worked out, then, that *Voyager 1* would travel on a slightly shorter path to Jupiter and Saturn on the trajectory designed by the navigation team. So even though it launched three weeks later, *Voyager 1* passed *Voyager 2* by the time both spacecraft were passing through the Main Asteroid Belt between Mars and Jupiter.

Building and launching a spaceship is just the beginning of the work of the human team back on Earth. There has to be a way to track them to make sure they're heading in the right direction, to steer them if they need course corrections, and of course to communicate with them—sending them commands needed to perform their mission and getting back the photos and other data that they were sent to collect. That critical communications job is the work of the men and women of NASA's Deep Space Network (DSN), a trio of giant radio telescope facilities in California, Australia, and Spain that is managed by JPL. The DSN's sensitive antennas are spread roughly equally around the Earth so that at least one of them can always be in contact with any of the thirty or so active space missions being run by NASA and other space agencies. These radio antennas and their diligent operators are booked solid twenty-four hours a day, seven days a week, keeping tabs on the trajectories of all these spacecraft, sending them routine commands or responding to "spacecraft emergencies" that they sometimes have, and receiving and relating the billions of bits of digital data that are relayed back to the Earth every day to operations centers around the world, like JPL.

I visited the DSN station in Canberra, Australia, a few years ago and stood in awe under the superstructure of the 70-meter-wide (more than 200 feet) antenna used to communicate with *Voyager* and other missions. The DSN's radio telescopes need to be so big because the signals from *Voyager* are so small. By the time the spacecraft got out to Jupiter, for example, *Voyager*'s 23-watt radio transmitter produced a signal that was only about a hundred-millionth as powerful as a cell phone battery by the time it reached Earth. These days, with both spacecraft now well beyond the orbit of Neptune, the power levels received at Earth of *Voyager*'s radios are more than five hundred times fainter than they were at Jupiter.

While the DSN sends command sequences to the spacecraft, the people making those commands, people called *sequencers*, work at operations centers like JPL or at other government labs or universities around the world. Sequencers are like the accountants of the space business. All they do is figure out how to get complex machines to do intricate things with small, simple sets of instructions, often written in arcane languages. For a spacecraft, a *sequence* is like a time-stamped to-do list of individual commands. Fly to a certain place, turn on the camera, point it in that direction, take twelve photos, turn off the camera, turn on the magnetic-field sensor and collect those measurements for twenty hours, restart the camera, point it at another location, and so on and so forth. Sequences are not written in English, though, but in computer code that ultimately has to be broken down into the binary strings of ones and zeros that can be transmitted to the spacecraft by the DSN.

This is hairy, intricate work, with catastrophic (for the mission) consequences for making mistakes, and so sequencing often attracts certain kinds of fastidious, patient, perhaps even borderline OCD kinds of people. Perhaps you know people who can quickly spot a typo on a newspaper page, who are masters of details that others might consider trivial, who can easily visualize and describe the three-dimensional positions of objects in their heads, or who can easily see patterns in large groups of numbers—these are the kinds of highly sought-after traits in sequencers. Even though fault protection against fatal mistakes is often built in to the software that spacecraft run, scientists and engineers learn to *trust* sequencers with the very life of the mission. This is especially true of flyby missions like *Voyager*, where all the action can happen only once, as it zips past each of the planets at high speed.

My planetary science colleague Candice ("Candy") Hansen started her career at JPL as one of these sequencers, working with the *Voyager* imaging team. Her job title was "experiment representative," which meant that she was a liaison between the rank-and-file scientists on the team—many of whom didn't know much about how the cameras or the spacecraft were operated—and the rank-and-file instrument and sequence engineers on the team—many of whom, such as a fresh-out-of-college engineer named Suzy Dodd, didn't know much about the kinds of science that *Voyager* was being asked to do. Candy spent most of her time designing the detailed plans for pointing *Voyager*'s cameras at specific targets of interest (especially icy moons), devising the best ways of stringing individual photos into larger mosaics. Another of Candy's jobs was to try to estimate the exposure times for *Voyager*'s photos of planets and moons. In some cases, the team would be photographing places that had never been seen before, and so they had no idea if the surface would be bright like snow or dark like coal (or, like some places turned out to be, both). If she commanded exposure times that were too long, the photos would come out all white ("saturated") or super blurry because of *Voyager*'s high speed relative to the targets of interest. If she commanded exposure times that were too short, everything would come out all black or at least way too dark to be able to be scientifically useful. The stakes were high, given that each spacecraft had only one shot to photograph these places as it sped past.

Someone on the team like Candy would be tasked with making the initial estimates of the exposure times and other parameters for each of *Voyager*'s planned images. They'd use whatever information they could, from ground-based telescope observations, laboratory studies, theoretical calculations, *Pioneer* data, or previous *Voyager* images (especially for *Voyager 2*, given that it passed through the Jupiter and Saturn systems after *Voyager 1* had gone through first). Then they would present their estimates to the full team for review and critique. They'd make changes, rerun calculations, redo the imaging sequences, and present it again. And sometimes again, and again. Each time, they'd have to run their plan by one of the sequence engineers, like Suzy Dodd.

"They would design the science observations for each of their instruments," Suzy recalls, "and then we would lay them out and see if they would fit within the resources on the spacecraft."

Pressure and stress levels mounted as changes were being made closer and closer to the deadline for uplinking the final imaging sequence to the spacecraft. To help relieve some of her own stress, and the team's, Candy would bake cookies. "The later the change in the sequence, the more cookies I would bake for the sequence team to thank them for letting us make the *really* late changes," she recalls. During the pressure-packed lead-up to each of *Voyager*'s planetary encounters, Candy was hardly ever at home. If she was, she was baking. . . .

Despite the many checks and double checks, by multiple people, "Mistakes still happened," Candy says, "and you never forget them. Estimating exposure times was a big deal. For example, we were updating Neptune exposures in the last iteration of the sequence before it went to *Voyager 2* and I miscalculated the exposure times on two images. I was working with a specific science team member on the planning of these images, and I don't think he ever forgave me . . . and here it is, twenty-five years later, and I still feel bad about that."

But the vast majority of time, things went right. Linda Spilker, another instrument scientist with *Voyager*, and now the JPL project scientist for the *Cassini* Saturn orbiter, recalled the personal rewards from seeing the process run well from end to end. "I found tremendous satisfaction in seeing the data from an observation that I had carefully planned with input from the IRIS team reach the ground and be carefully analyzed," she says, referring to her role in planning observations from *Voyager*'s infrared radiometer interferometer and spectrometer instrument (IRIS). "What new revelations would that data contain about mighty Titan or intriguing Triton? My job was to carefully review each and every spacecraft command for every *Voyager* IRIS sequence and to make sure that those commands were right." High pressure, for sure, but high rewards, too.

The kinds of job-related stresses that sequencers faced during the one-shot flyby missions of *Voyager* are somewhat different from the stresses faced by sequencers and their teammates operating orbital spacecraft, such as the *Mars Reconnaissance Orbiter* or the Saturn orbiter *Cassini* (both of which Candy is currently involved with), or landed missions like the Mars rovers that I have worked on. In those cases, if an exposure time is botched or some other sequence error occurs, it's often (though not always) possible to take a "do-over" the next day, or the next orbit, to get it right. But still, in my experience and in the experience of almost everyone I know in the business, such mistakes are rare. The men and women tasked with doing the often thankless, almost anonymous, work of the nuts-and-bolts day-to-day care and feeding of spacecraft—including developing the detailed, step-by-step sequences that tell our far-flung robots exactly what to do—are among the most careful, thoughtful, conscientious, and trustworthy people I know. Dozens of people like Candy Hansen, Linda Spilker, and Suzy Dodd filled such roles on *Voyager*. And they are only a subset of the many thousands of people that it typically takes to design, build, test, launch, and run a modern ship of exploration. Charley Kohlhase estimated that a total of 11,000 work years were devoted to the *Voyager* project through the Neptune encounter in 1989—equivalent to a third of the labor force needed to complete the Great Pyramid at Giza for King Cheops.

Working closely in the trenches with science and engineering and management colleagues on a high-stakes, high-stress space mission is a bonding experience for such teams of people. I can imagine that the experience is perhaps comparable in some ways (though not life threatening!) to what it must feel like for a team of mountaineers to tackle a particularly dangerous and challenging peak together. You learn to rely on your teammates, to understand and respect the entire range of skills and capabilities that it took to get to the top, and to celebrate successes and mourn failures as the life-changing events that they are.

"It was a wonderful, wonderful, wonderful thing to be involved with," Candy Hansen says, reflecting back on her *Voyager* years. "It was like working on a common cause, a common passion. You were working with people who really wanted the same goal, the same outcome."

"We were young and fun!" recalls Suzy Dodd (who rose up through the ranks to eventually become the project manager of *Voyager* in 2010), while lamenting also, almost under her breath, "and now we're old and gray. . . ."

When I asked Charley Kohlhase to try to pinpoint the glue that helped to bind the team together through all the years and the billions of miles traveled, he says the key was that "the *Voyager* team had tremendous devotion to the job." Perhaps *Voyager* sequencer Ann Harch, now a veteran of many more NASA missions over the past thirty years, summed it up best: "I can say without reservation that *Voyager* was the best-managed mission I ever worked on. The managers trusted their team chiefs and let them do their jobs. There was no micromanaging. Miracles happened with that kind of management. It truly was a magical mystery ride!"

3

Message in a Bottle

DURING THE LAST few centuries, humans have shown a remarkable ability for decoding messages in languages or codes that they had never previously encountered. Linguists were able to decipher the Greek written language Linear B from about BCE 1450 without any ancient Greeks around to provide tips. During World War II, Cambridge mathematician Alan Turing and the Allied Forces were able to decipher the ingenious Enigma machine ciphers used by the Nazis to great effect in North Atlantic naval battles. It seemed reasonable, then, to assume in the 1970s that any form of intelligent alien life as smart or smarter than ourselves would be able to decipher a message we sent to them, no matter how rooted it was in our culture, solar system, and galactic address. After all, we would be trying to make ourselves understood.

But then, if you were given the chance to compose a message to the future, to some conceivable or even inconceivable intelligent life out there in our galaxy, what would it be? It is a harder problem than it might first appear. . . .

That was the precise opportunity that presented itself when a small group of visionaries realized that long after the completion of *Voyager*'s scientific mission, the two spacecraft would continue to travel on silently, headed on an irreversible course out of our solar system and into the uncharted vastness of space. When the trajectories for *Voyager 1* and *Voyager 2* were chosen early in the history of the project, their ultimate long-term fates were sealed: both spacecraft would be traveling so fast because of their gravitational slingshots past the giant planets that they would achieve escape velocity from the solar system. That is, they would no longer be in orbit around the sun, but instead would be on one-way trips to interstellar space, no longer bound to their parent star like the rest of us. The *Voyagers* would be emissaries—human artifacts, time capsules of a sort, technological snapshots of what our species and our civilization was capable of doing during the time when the spacecraft were built and launched. The idea of including a message in those bottles cast into the cosmic sea seemed appropriate.

But what message would be the right one to send? This weighty question would ultimately be pondered by a small group of scientists, writers, and artists led by Carl Sagan—a group who would also reach out for advice and input to a larger range of scientists, artists, philosophers, teachers, and dignitaries throughout the world. Could people rise to the challenge of crafting a unified message representing not their own parochial interests or agendas, but their hopes, dreams, and experiences as citizens of Planet Earth? Sagan believed they could. The secretary-general of the United Nations at the time, Kurt Waldheim, proposed (unsolicited by anyone involved with *Voyager*) a moving letter as his contribution to the message, with the words, "We step out of our Solar System into the universe seeking only peace and friendship, to teach if we are called upon, to be taught if we are fortunate." With this and countless other words, music, and images that emerged from the challenge, Sagan and others would try to bottle our dreams and send them adrift with our aspirations, etching them into a golden time capsule known as the *Voyager* interstellar message.

Carl Sagan had been largely responsible for an interstellar message in the form of a plaque sent off with the NASA *Pioneer 10* and *Pioneer 11* spacecraft. The *Pioneers* were humanity's first missions beyond Mars, the first to fly past Jupiter (in 1973 and 1974), the first to fly past Saturn (*Pioneer 11* in 1979), and the first human-made objects to be accelerated beyond the escape velocity of the sun. They were pathfinders for the follow-on *Voyager* missions, demonstrating many of the technologies and celestial navigation methods that would later prove critical to the *Voyagers'* successes; for example, demonstrating the use of gravity assist at Jupiter, and proving that spacecraft could pass unscathed through both the Main Asteroid Belt and the plane of Saturn's rings. In addition, the *Pioneers* did some of the initial science scouting that would allow the *Voyager* science instruments to be optimized; for example, measuring the powerful radiation and magnetic-field levels at Jupiter and Saturn, and taking some of the first high-resolution images of those planets.

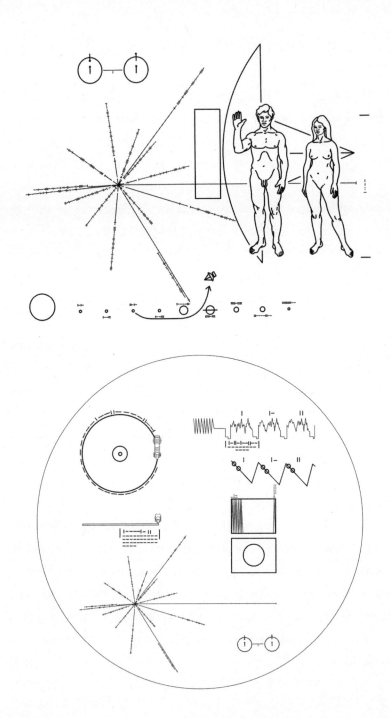

EXPLANATION OF RECORDING COVER DIAGRAM

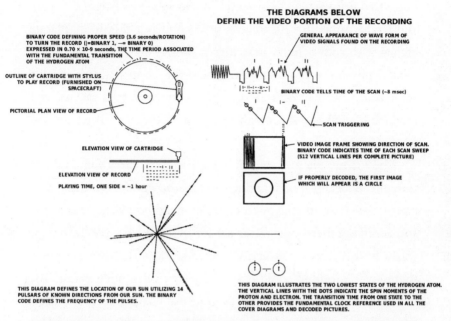

Messages from *Pioneer* and *Voyager*. *Facing page, top:* Plaque designed by Sagan, Drake, and Salzman Sagan for the *Pioneer* missions. *(NASA/JPL) Facing page, bottom:* Plaque designed by Sagan, Drake, Lomberg, and others for the *Voyager* missions, and engraved onto the cover of the *Voyager* Golden Record. *(NASA/JPL) Above:* Explanation of the symbols and markings used on the *Voyager* plaque. *(NASA/JPL)*

In the early 1970s, Sagan; his wife, the artist and writer Linda Salzman Sagan; and the pioneering Search for Extraterrestrial Intelligence (SETI) astronomer Frank Drake acted on the idea put forward by science journalist Eric Burgess and author Richard Hoagland that humanity shouldn't miss the opportunity to include a message of some kind on these high-tech emissaries that were about to be cast out forever into interstellar space. With access to *Pioneer* project officials for the technical details of the spacecraft and to NASA headquarters officials for the required permissions, but with only three weeks to get the job done, Sagan, Salzman Sagan, and Drake came up with a clever gold-anodized aluminum plaque etched with rudimentary drawings and markings based on fundamental physics and astronomy, which they hoped would enable some comparably (or more) intelligent extraterrestrial species who might intercept the spacecraft in the far future to tell where and when it came from.

The *Pioneer* plaques were intended only to "carry some indication of the locale, epoch, and nature" of the builders of the spacecraft rather than to convey to our alien neighbors *why* we sent the spacecraft out there, or what we are (or were) like as a civilization. It was a hastily conceived and executed project that the designers themselves admitted could potentially have been done better. The message made some assumptions in its use of symbols (such as arrows) that may make sense only from a human cultural perspective (we long relied on arrows and spears and know which way they travel), and also assumed that intelligent aliens would have at least a rudimentary knowledge of basic physics and chemistry. Some of the references to our location relative to nearby rapidly spinning neutron stars (pulsars) might not make sense to beings in other parts of the galaxy, or in the far future when those pulsars have changed. It's even been criticized by some as interstellar pornography because of its depiction of a naked man and woman.

Be that as it may, it is humbling to realize, as the plaque makers pointed out, that given the extremely low rates of erosion in deep space, "Pioneer 10 and any etched metal message aboard it are likely to survive for much longer periods than any of the works of man on

Earth." In their 1972 description of the *Pioneer* plaques, Carl Sagan and colleagues noted that "the message can be improved upon, and we hope that future spacecraft launched beyond the solar system will carry such improved messages."

THE GREATEST CONCEPT ALBUM EVER

Indeed, that chance came just a few years later, with the development of the *Voyager* missions. In December 1976, the *Voyager* project manager at the time, John Casani, with the enthusiastic support of Project Scientist Ed Stone, gave Carl Sagan the go-ahead to organize and lead the effort to place messages on the two *Voyager* spacecraft. Sagan happily accepted and set about consulting experts of every sort about how this message should be constructed. He spoke with astronomers, physicists, biologists, science-fiction writers, and philosophers. The initial thought was that this would be an expansion on the ideas developed for the *Pioneer* plaque. However, the course of this project was forever changed when, in January of 1977, Sagan's Cornell University colleague and fellow astronomer Frank Drake suggested sending a long-playing phonograph record (or LP as they were known) instead. At that moment, the Golden Record was born.

Frank Drake is widely known and respected as one of the pioneers in the development of SETI and its serious search for extraterrestrial life. An astrophysicist and radio electronics expert by training, Drake used a powerful radio telescope to beam short, friendly, and hopefully understandable greetings to a number of nearby stars in 1960's Project Ozma. He conjured up the famous

"Drake equation"—a serious attempt to mathematically estimate the number of intelligent civilizations in our galaxy by stringing together a bunch of probabilities (many still not known or even very well constrained) regarding, for example, how many stars have planets, how many planets have life, whether that life would become technological, and if so, how long that life might survive. I wonder if Drake's idea for putting an LP on *Voyager* was influenced by the many "concept albums" that were climbing the US music charts in the early to mid-1970s. Musical dramas by Pink Floyd, ELO, Styx, David Bowie, and many others certainly could have helped set the context for the *Voyager* record team pulling together what could end up being the greatest concept album of all time.

The idea of the record was instantly appealing to Sagan and others because of the relative permanence that could be attained by etching information—in this case sound and digital representations of pictures—into grooves in the disk that could last for billions of years. The beauty of the record was that it would be possible to send not only pictures and information but music, historical as well as contemporary. It would be an opportunity to send something of the human spirit along with the more straightforward (yet still challenging to interpret) messages of human intelligence. Most of us would agree that music can capture human emotions to a degree beyond anything that we can convey with equations. It was an opportunity to transmit feelings that could just, maybe, be comprehended by advanced life-forms from another time and place. As the musicologist Robert Brown, then of the Center for World Music in Berkeley, California, one of the key individuals who was deeply involved with the selection of music to be sent to the stars, noted, "If we don't send things we passionately care for, why send them at all?"

Sagan and others discussed the possibility of also sending works by the great artists throughout human history, but ultimately decided not to because, according to space artist and *Voyager* Golden Record design team member Jon Lomberg, "To fairly depict the range of human art was a task of curation beyond the ability of the picture group to complete in time. Better not to do it than to do it badly." It is also easy to imagine that it would be difficult enough for alien beings to understand our photographs, so abstract or stylized paintings of our world would have even less of a chance of being comprehended. Still, some would have liked to see representations of Van Gogh's painting *Starry Night*, Michelangelo's sculpture of David, or Hokusai's woodcut of *The Great Wave off Kanagawa* included, no doubt, but Sagan and the others involved made their call.

Everyone involved in crafting the *Voyager* interstellar message realized, of course, that the chances of any message sent into space in this manner being actually *received and interpreted* by other beings would be extremely tiny. I've made the comparison to a message in a bottle thrown into the ocean. But "ocean" isn't close to the right scale. The seemingly limitless emptiness of space is difficult for us to comprehend. Carl Sagan, when talking about the odds of the *Voyagers* encountering other civilizations, would imagine affixing a small number of balloons randomly to the walls inside a sports arena like Madison Square Garden, and then having someone randomly throw darts at the walls. There would be a chance that they could hit and break one of those balloons, but the chance would be staggeringly small. Nevertheless, for *Pioneer*, and again for *Voyager*, the idea of creating and sending an interstellar message that *could* be received by someone or something was pursued with gusto. It was ultimately an opportunity to reflect on what we had to offer other worlds, other civilizations. What a testament to the human spirit, and to our perseverance in spite of all bets being against us.

Still, not everyone was thrilled about the chance to send a message to outer space. Some, in fact, thought it was a very bad idea indeed. We would be shouting out, "Hello! Here we are. Come and get us. And by the way, here is a map!" There were those who perceived a grave danger in this. What if the recipients were hunters? What if they were malevolent, or just plain hungry? This point of view had been espoused for years, especially in relation to the SETI projects that aimed to communicate with distant extraterrestrial life-forms by means of radio waves. In 1974, scientists at SETI sent out the first message directly seeking contact with extraterrestrial life, aiming the Arecibo radio telescope at M13, a "nearby" globular cluster of stars just 25,000 light-years away. While there was great support for the project and almost entirely positive reactions from scientists as well as the public, one quite serious objection was made by Sir Martin Ryle, a Nobel laureate astronomer no less, who viewed this as a dangerous business, us foolishly calling out to potential enemies and announcing our location. He even attempted to put some rules in place that would forbid similar attempts at communication in the future.

However, the culture of the *Voyager* project was in general far more optimistic and unafraid. They celebrated contributions such as the one an Indian sent in Rajastani: "Hello to everyone. We are happy here and you be happy there."

Today the debate about the wisdom of attempting contact with intelligent life out there in the universe rages on. In the past few decades, more radio messages have been sent out, such as the "Cosmic Call" messages—two interstellar radio messages sent to a variety of nearby stars in 1999 and 2003. And with the actual recent discovery of a multitude of planets around other stars (so-called exoplanets), a discovery enabled by our ever more sophisticated astronomical instrumentation, we have more reason than ever to believe that we are not alone. Another of the most notable proponents of keeping our cosmic mouths shut is Stephen Hawking, Cambridge theoretical physicist, cosmologist, and one of the great thinkers of our time. According to Hawking, we are simply not evolved enough to make such contact. To make his point, he uses the analogy of the arrival of Christopher Columbus in the Americas, "which didn't turn out very well for the Native Americans." He went on to say, "We only have to look at ourselves to see how intelligent life might develop into something we wouldn't want to meet."

Regardless of the arguments on both sides, contemplating the pros and cons of this problem may be pointless, for there are many who believe that it is already too late for this debate. We send electromagnetic waves out into space all the time, announcing our location and the vagaries of our cultures, in the form of TV broadcasts, powerful military radars, and communications with the spacecraft that are in operation by the world's space agencies. The makers of the *Voyager* message decided that, whether we like it or not, we continue to say, day after day, "We are here." If we are broadcasting our presence already anyway, why not take some care to make some of what we send be intentional messages? Don't just babble, say something.

Choosing the sounds and images to include on the Golden

Record was no simple task. How would you choose to represent our planet in a couple dozen songs and a bit more than 100 low-resolution images? Would you send some kind of sales pitch, or a neutral sample of artifacts? Would you decide to focus on the uplifting and admirable side of humanity, or would you opt for a more balanced depiction, putting forth our greatest achievements alongside the ever-so-brutal moments that are also a sad but undeniable part of our world's history? Carl Sagan decided to leave out some topics that represent the weakest side of human nature—topics like famine, disease, injustice, and war. While the dark side of humanity cannot be denied, it is not the part of us that the Golden Record team wanted to send out into the stars. If the messages aboard the *Voyagers* ended up being the last surviving artifacts of our world, they would signify the brighter side of human nature. After all, *Voyager* folks, and I count myself among them, wanted to send out signs of our hopes, not our regrets.

There was also a more practical side to the decision to avoid topics like war: Sagan and others in the group didn't want what we sent to be perceived as a threat. What sort of a message does the picture of, for example, a mushroom cloud from a nuclear explosion send? What if the recipients were to think that we were the aggressors and *we* were after *them*? Best to leave them out.

"But then again, Carl was a subtle thinker and he thought on many levels," my friend Jon Lomberg told me recently. "Remember, this was still in the '70s. Then, like now, there was a lot of political polarization in the country, and I think it was important for him not to politicize the Record in any way. If he started showing pictures of certain atrocities, then people who suffered some other atrocities could seek equal time to highlight their plight. It was a road he didn't want to go down. Carl wanted it to have exactly the kind of positive influence that it did. It was like asking ourselves, 'Could we measure up to the *Voyager* Record?' This is us at our best, or at least not at our worst. This approach made the record aspirational, especially because I don't think any of us cherished a real hope that it would ever be found. So the only audience we know about is the audience on Earth."

The making of the record and all the thoughtful choices that were made during the whirlwind six-week project are described in great detail in the book *Murmurs of Earth,* written by the team itself—astronomers Sagan and Frank Drake, and artists and writers Jon Lomberg, Timothy Ferris, Linda Salzman Sagan, and Ann Druyan. Many others contributed in important ways as well, and they are given their due credit in the pages of the book. The collaboration within the team was intense and close, given the time frame in which they had to produce a final record—just six weeks from approval to final product. Surely that must have kept them up at night.

Jon Lomberg says, "Well, they'd had the experience of having only three weeks for the *Pioneer* plaque, so that was a start. But I think, and this is just a guess, that Carl knew, perhaps correctly, that for this not to get bogged down in congressional oversight—or who knows how many other people wanting to put their two cents in—that if it were done sort of at the last minute and sort of as a fait accompli, then people wouldn't really have the chance to say we shouldn't do it."

It makes sense. This is, in fact, the same philosophy that Jon and

I, along with Steve Squyres, Bill Nye, and other colleagues, had taken when we devised the design, messages, and other "furniture" that turned an esoteric camera calibration target on the Mars rovers *Spirit*, *Opportunity*, and *Curiosity* into Martian sundials, or "MarsDials." The idea was to be able to calibrate the cameras using swatches of colored and gray-scale materials, but the bigger-picture idea was also to help teach kids about timekeeping and understanding our place in space using only sticks and shadows—much like the third-century BCE Greek mathematician and astronomer Eratosthenes had done to accurately estimate the size of our planet. We figured, apparently as Carl Sagan did for the *Voyager* Golden Record, let's keep this under the radar, lest it get killed by committee.

The *Voyager*'s two-sided gold-anodized copper LP contains an hour and a half of music (27 pieces in all), 116 digitized photographs, and a catalogue of terrestrial sounds (such as the chirping of crickets) and voices (such as short greetings in fifty-five languages, including a "hello from the children of Planet Earth" in English from Carl Sagan's six-year-old son, Nick). The record was housed inside a circular gold-plated aluminum casing to keep it protected from radiation, which could slowly weaken and corrode the metal, and from erosion by high-speed micrometeorites, which could more quickly pit and gouge it. A stylus (needle) and its cartridge are included nearby, along with other information and instructions on the outside of the case for how to play the record. With the vinyl-music era now fading into history, it is perhaps ironic that not just intelligent aliens but also most listeners of popular music today would require those instructions.

The number of musical pieces and pictures that could be

included was limited by the necessity of preserving the fidelity of the sounds and images while digitizing them to all fit on one double-sided record. Sagan and Ferris took charge of arranging the musical selections, Druyan was responsible for the sounds of Earth (all sounds outside of the musical selections), Salzman Sagan took charge of the multilanguage greetings, and Drake and Lomberg assembled the image collection, which consisted of digitized photographs and drawings, some of which were created by Lomberg himself when the images the team was looking for could not otherwise be found.

The 116 pictures are divided into two categories: photographs, and diagrams intended to teach the recipients how to comprehend the photographs. The photos were chosen with a single purpose: to convey information that is unique to our planet and its people. If the photos happened to also be artful, that was considered a bonus. The diagrams were there to provide information, mostly scientific and mathematical, that would explain things like the composition of the air we breathe here on Earth. In fact, the first diagrams that our future extraterrestrial friends would encounter were etched into the cover of the record case itself. The Golden Record was mounted on the front of the spacecraft, shielded by its gold-coated aluminum cover, thirty-thousandths of an inch thick. On this cover are not just the instructions on how to play the record but also how to reconstruct the pictures encoded on the record (using the vibrations of the record needle as proxies for the brightness of each digitally reconstructed pixel), and a map indicating where and when this message was launched ("Earth, 1977"). But what is the best means of conveying this information to beings who would not be likely, by

Unlocking the Code. Examples of some of the 116 images encoded onto the *Voyager* Golden Record.

•	= |	= 1	||-- = 12
••	= |-	= 2	||--- = 24
•••	= ||	= 3	||--|-- = 100 = 10^2
••••	= |--	= 4	|||||-|--- = 1000 = 10^3
•••••	= |-|	= 5	
••••••	= ||-	= 6	
	|||	= 7	
	|---	= 8	
	|--|	= 9	
	|-|-	= 10	

$$2+3=5$$
$$8+17=25 \qquad 5+\frac{2}{3}=5\frac{2}{3}$$
$$\frac{1}{2}+\frac{1}{3}=\frac{5}{6} \qquad 2 \times 3 = 6$$
$$\frac{1}{3}+\frac{1}{5}=\frac{8}{15} \qquad 13 \times 28 = 364$$

Image 3, defining the mathematical symbols and numbers used elsewhere among the images. (*Frank Drake*)

$$1\frac{42}{100} \times 10^9 \underline{t} = 1\underline{s}$$
$$86400\underline{s} = 1\underline{d}$$
$$365\underline{d} = 1\underline{y}$$
$$6 \times 10^{23}\,\underline{M} = 1\underline{g}$$
$$1000\underline{g} = 1\,\underline{kg}$$
$$6 \times 10^{27}\,\underline{g} = 1\,\underline{e}$$

$$\frac{1}{21}\underline{L} = 1\,\underline{cm}$$
$$1\underline{L} = 21 \times 10^8\,\underline{\mathring{a}}$$
$$10^2\,\underline{cm} = 1\,\underline{m}$$
$$1000\,\underline{m} = 1\,\underline{km}$$

Image 4, providing a visual definition of the basic units of length, mass, and time used among the images, using the fundamental properties of hydrogen as the basis. (*Frank Drake*)

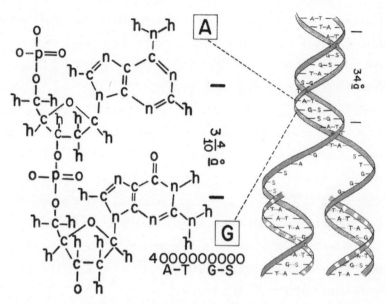

Image 15, showing the basic chemical and physical structure of DNA molecules, the building blocks of all life on Earth. (*Jon Lomberg*)

any stretch of the imagination, to understand human languages or conventions? This is where people like Frank Drake were indispensable. "Frank Drake is a bit like Thomas Edison and Albert Einstein rolled into one," marvels Jon Lomberg.

Having been deeply involved in preparing earlier interstellar messages as part of the SETI program, as well as the plaques on the *Pioneer* spacecraft, Drake and others drew heavily from their past work. As on *Pioneer*, the location of our home planet in the solar system was indicated on the gold-anodized record case using a map whereby one can orient oneself using the location relative to a number of prominent pulsars—rapidly rotating neutron stars each with its own distinct frequency. And furthermore, since the frequency of these pulsars changes slightly over time, not only our *location* but our *time* could be conveyed. But how would the pulsar frequencies be represented? That question led to another diagram etched on the cover of the *Voyager* record (also drawn from *Pioneer*): a diagram of the hydrogen atom (see figure on page 86).

Hydrogen is the most abundant element in the universe, and it is rather simple—just one proton and one electron. The idea was to use the time it takes for the hydrogen atom to transition between its two lowest-energy states as the fundamental time scale (0.70 billionths of a second) and this would be the basis for how all other time would be indicated on the record. The pulsar frequencies could be represented in multiples of "hydrogen time." Since hydrogen is so basic, it was assumed that intelligent life elsewhere would be able to decode this message. All numbers were represented in binary, a simple means of counting whereby all numbers can be represented by combinations of zeros and ones. So on the cover of the Golden Record, there is the pulsar map with frequencies represented in binary. The instructions for playing the record tell the user how fast to spin it (which happens to be 3.6 seconds per rotation, half the speed of what was then the standard for a 33⅓ rpm vinyl LP) and also provide guidance on the more complex procedure of reconstructing the pictures, which must be traced with a series of interlacing vertical lines. As a sort of a test for anyone attempting to decode the pictures, below these instructions there appears a picture containing only a circle. This is the image that would appear if the first picture in the set were decoded properly. But if it came out as an oval, for example, instead of a circle, that would be an indication that something must be corrected in the decoding procedure. In this sense the circle was a calibration picture.

All of this was very clever, but who knows whether intelligent life originating elsewhere in the galaxy would have the correct faculties to decode it. Sagan, Drake, Lomberg, and the other team members presumed that any aliens who found the record would understand atomic physics and recognize hydrogen as the universe's most common element, and perhaps would also recognize other molecules as common (H_2O, DNA?). They presumed that the aliens would use vision and hearing of some kind to reconstruct and understand the music and pictures, although one can imagine purely tactile ways to perceive the content as well. They presumed that the finders would have an understanding of deductive reasoning, of representations of time, and of the causality of events. Perhaps even the idea of pictures and music being used to tell a story would end up being truly alien to them, however. "It's wise to try every possible approach," Frank Drake wrote in a recent interview, "because we're not very good at psyching out what extraterrestrials might actually be doing."

If the exercise were repeated today, perhaps we would include more sophisticated representations of physics, chemistry, astronomy, and biology, taking advantage of forty years of technological advances since the 1970s. Maybe Madonna or Michael Jackson or Lady Gaga would find their way into the recordings. Maybe we would find a way to capture more of the complexity and dichotomy of our species (being capable of both such beauty and such horrors) without necessarily appearing as threatening. Perhaps we would embrace more of the historical honesty and social introspection characteristic of our times by telling the finders, effectively, "Hey, check us out a little more if you can and make sure that you're not going to be sorry you called." A little of Hawking's medicine might not be such a bad idea. If not, then perhaps at least today's lawyers would want us to plaster a big CAVEAT EMPTOR on the front casing to help stave off lawsuits in case we don't meet alien expectations.

NASA did impose bureaucratic hurdles in the approval process for the contents of the record—Sagan's cohort was not completely under the radar. For example, although the somewhat contentious cartoonlike drawings from *Pioneer* of nude humans were used once more on *Voyager*, an actual (tasteful) photograph of a nude pregnant woman and a nude man holding hands was left out for fear of an adverse public reaction.

In addition to NASA executive committee vetoes, in some cases copyright permission was denied or its pursuit simply abandoned. Because this record did not actually promise any sales, there was little incentive for any corporate entity, particularly a recording company, to accelerate the approval process. I asked Jon Lomberg if there was anything missing from the *Voyager* record for this kind of reason. "The Beatles," he responded instantly. All four members of the band wanted "Here Comes the Sun" included—but their publisher wouldn't grant the rights.

"In some ways the Beatles were the most obvious choice to include on the music. They were still at the peak of their fame, even though they'd broken up five years before. It would have been like putting on Shakespeare—who is going to seriously say that Shakespeare doesn't belong among the greatest hits of Earth's literature? The Beatles were sort of the absolute peak of Western musical achievement at the time. So that was a big disappointment. It made the suggestion of who to replace them with by no means obvious, because there was a whole tier of great musical performers. In the end, Tim Ferris's choice of Chuck Berry was a good solution."

Bill Nye recalls, "I was in class at Cornell in the spring of 1977 when Carl Sagan asked us which Chuck Berry song to put on the records. He actually pitched 'Roll Over Beethoven,' but we all insisted that the record include 'Johnny B. Goode' instead. And so it came to pass." Passing the youth test, it apparently passed the social-media test of the day as well: Steve Martin did a skit on *Saturday Night Live* in spring 1978 where an alien's response to the *Voyager* record was "Send more Chuck Berry!" "It is a sobering thought, though," Jon Lomberg nonetheless laments, "that it was easier to send a record into deep space than it was to try to market it here on Earth."

Drake, Lomberg, and the others searched through picture books in libraries, in magazines from *National Geographic* to *Sports Illustrated*, and in NASA's photo services. When they couldn't find a photo they were looking for, the picture group composed and shot some of their own. Six, in fact. Lomberg also devised a dozen original diagrams that contained important visual and other information. He would later go on to use his skills to create much of the graphic space art for Carl Sagan's original 1980 television series, *Cosmos*. They worked hard to try to choose images that they expected would be the most informative and the least confusing. And they spent a lot of time playing extraterrestrial head games—trying to put themselves in the shoes (or whatever) of beings viewing the pictures without the benefit of the unconscious context that we all have by virtue of living on Planet Earth.

Upon decoding the Golden Record, the first image that would pop out would be the calibration circle. After that, it would be time to learn a bit of our scientific language. Images were constructed to show how our numbers work and to define things like distance and mass in terms of universally (in the true sense, it was hoped) known quantities like the mass of the hydrogen atom and the wavelengths of radiation emitted when hydrogen changes energy states. In that way, it would later be possible to convey more complex ideas, such as how many masses of hydrogen atoms make up the mass of a human being (it's a huge number with twenty zeros at the end). Providing this dictionary of sorts was the key to labeling many of the photographs in meaningful ways, such as marking pictures of all of the planets in our solar system with their diameters in kilometers.

In keeping with the idea of sending pictures that are somehow unique to Earth, the group decided that they would have to include diagrams showing the structure of DNA. After all, it is the recipe for all life on Earth. It is fascinating to ponder whether all life must necessarily use the same building blocks, and whether DNA is universal. This question is a great driving force behind the current push by the astronomy, planetary science, and nascent astrobiology communities to develop missions to search for evidence of life, or at least evidence of habitable environments, in our own solar system. Without any examples beyond Earth, such questions are presently not much more than speculation. The DNA images would certainly convey our understanding of our brand of life, but possibly also our inability to yet grasp our place in the galaxy and beyond.

A small number of photographs of other planets in our solar system were included to show our "neighborhood," but by far the bulk of the photographs were meant to depict the variety and uniqueness of life on Earth. These included pictures of trees, flowers, animals, seascapes, and mountains, as well as those depicting the many cultures and behaviors found across the globe. Everything from how we eat (there is a picture of a man eating grapes in a field and another of a woman eating grapes in a supermarket) to how we learn, dress, dance, run, socialize, and build is represented. Some of the great architectural and engineering accomplishments of our planet are represented by the Sydney Opera House, an airport, a radio telescope, and the Taj Mahal. The way our bodies move is depicted by a stroboscopic picture of a gymnast performing her routine on the balance beam. A time scale of five seconds is thoughtfully included on this photo to announce that we move around in a matter of seconds as opposed to milliseconds or years, for example. This level of detail is typical of the concerted mental effort that was devoted to choosing each photograph. Even a shimmering picture of a sunset over the ocean—presumably chosen simply for its beauty—had an educational component to it, since an educated alien with an understanding of the physics of optics and fluid mechanics could use the colors and patterns in the sky and the ocean to deduce some aspects of the pressure, chemical composition, and other properties of our planet's atmosphere and oceans. Regardless of whether any future decoders of the Golden Record would infer all the layers of information intended, it is fun and at the same time mind-blowing to imagine what they might conclude about our home world when armed with only this limited set of pictures and sounds.

The music on the Golden Record, on the other hand, was less of a cerebral exercise. Who knew whether intelligent beings from across the galaxy would have a means for experiencing music. Could they even hear? Nonetheless, music represents the emotional side

Photos and Diagrams on the *Voyager* Golden Record

1. Calibration circle
2. Solar location map
3. Mathematical definitions
4. Physical unit definitions
5. Solar system parameters
6. Solar system parameters (continued)
7. The sun
8. Solar spectrum
9. Mercury
10. Mars
11. Jupiter
12. Earth
13. Egypt, the Red Sea, and the Nile
14. Chemical definitions
15. DNA structure
16. DNA structure magnified
17. Cells and cell division
18. Anatomy 1 (skeleton, front)
19. Anatomy 2 (internal organs, front)
20. Anatomy 3 (skeleton and muscles, back)
21. Anatomy 4 (internal organs, back)
22. Anatomy 5 (heart, lungs, kidneys, and main blood vessels, back)
23. Anatomy 6 (heart, lungs, kidneys, and main blood vessels, front)
24. Anatomy 7 (rib cage)
25. Anatomy 8 (muscles, front)
26. Human sex organs
27. Diagram of conception
28. Conception
29. Fertilized ovum
30. Fetus diagram
31. Fetus
32. Diagram of male and female
33. Birth
34. Nursing mother
35. Father and daughter (Malaysia)
36. Group of children
37. Diagram of family, with ages
38. Family portrait
39. Diagram of continental drift
40. Structure of Earth
41. Great Barrier Reef of Australia
42. Seashore
43. Snake River, Grand Tetons
44. Sand dunes
45. Monument Valley
46. Forest scene with mushrooms
47. Leaf
48. Fallen leaves
49. Snowflake over Sequoia
50. Tree with daffodils
51. Flying insect with flowers
52. Diagram of vertebrate evolution
53. Seashell (*Xancidae*)
54. Dolphins
55. School of fish
56. Tree toad
57. Crocodile
58. Eagle
59. Waterhole
60. Jane Goodall and chimps
61. Sketch of bushmen
62. Bushmen hunters
63. Man from Guatemala
64. Dancer from Bali
65. Andean girls
66. Thai craftsman
67. Elephant
68. Old man with beard and glasses (Turkey)
69. Old man with dog and flowers
70. Mountain climber
71. Gymnast
72. Sprinters
73. Schoolroom
74. Children with globe
75. Cotton harvest
76. Grape picker
77. Supermarket
78. Underwater scene with diver and fish
79. Fishing boat with nets
80. Cooking fish
81. Chinese dinner party
82. Demonstration of licking, eating, and drinking
83. Great Wall of China
84. House construction
85. Construction scene (Amish country)
86. House (Africa)
87. House (New England)
88. Modern house (Cloudcroft, NM)
89. House interior with artist and fire
90. Taj Mahal
91. English city (Oxford)
92. Boston
93. UN Building, day
94. UN Building, night
95. Sydney Opera House
96. Artisan with drill
97. Factory interior
98. Museum
99. X-ray of hand
100. Woman with microscope
101. Street scene, Asia (Pakistan)
102. Rush-hour traffic (Thailand)
103. Modern highway (Ithaca, NY)
104. Golden Gate Bridge
105. Train
106. Airplane in flight
107. Airport (Toronto)
108. Antarctic expedition
109. Radio telescope (Netherlands)
110. Radio telescope (Arecibo)
111. Page from Newton's *System of the World*
112. Astronaut in space
113. Titan *Centaur* launch
114. Sunset with birds
115. String quartet
116. Violin with music score

of humanity, and the pieces were chosen to convey the maximum feeling. Portraying the variety of human musical traditions was important, but when choosing a single piece from a culture, the emotion conveyed by the piece was paramount. The Golden Record's final musical repertoire is detailed in the table on pages 96 and 97. When looking over the list of musical selections chosen to represent our world, many will feel dissatisfied. After all, music is such an individual choice. Try explaining why you are moved by Beethoven's "Moonlight" Sonata (which I would like to have seen included, but I suppose the Fifth Symphony is also a decent choice). It is often impossible to articulate, but we can feel it in our hearts.

Many musical experts were consulted by Sagan and Ferris when choosing the final list, and the team's dedication to the requirement that the chosen music must touch the heart as well as the mind was striking in such scientific people. The decision to include multiple pieces by the same composer (three by Bach and two by Beethoven) is also interesting. In addition to the fact that Bach and Beethoven have produced some of humanity's finest works, it was thought by some, Lomberg included, that the inclusion of multiple pieces by the same composer would help illuminate our intent for the music, which was ultimately to convey mood and feeling. Even so, not everyone's personal favorites could be included. In addition to the omission of the Beatles, for example, Jon Lomberg was also disappointed that the music of Bob Marley was not represented. "At that time he didn't have the stature he did later, but still, his was real Third World music, and we needed more of that."

Music on the *Voyager* Golden Record

1. Bach, Brandenburg Concerto no. 2 in F, first movement, Munich Bach Orchestra, Karl Richter, conductor. 4:40

2. Java, court gamelan, "Kinds of Flowers," recorded by Robert Brown. 4:43

3. Senegal, percussion, recorded by Charles Duvelle. 2:08

4. Zaire, Pygmy girls' initiation song, recorded by Colin Turnbull. 0:56

5. Australia, Aborigine songs, "Morning Star" and "Devil Bird," recorded by Sandra LeBrun Holmes. 1:26

6. Mexico, "El Cascabel," performed by Lorenzo Barcelata and the Mariachi México. 3:14

7. "Johnny B. Goode," written and performed by Chuck Berry. 2:38

8. New Guinea, men's house song, recorded by Robert MacLennan. 1:20

9. Japan, shakuhachi, "Tsuru No Sugomori" (Crane's Nest), performed by Goro Yamaguchi. 4:51

10. Bach, "Gavotte en rondeaux" from the Partita no. 3 in E major for violin, performed by Arthur Grumiaux. 2:55

11. Mozart, *The Magic Flute*, aria no. 14, "Queen of the Night," Edda Moser, soprano. Bavarian State Opera, Munich, Wolfgang Sawallisch, conductor. 2:55

12. Georgian S.S.R., chorus, "Tchakrulo," collected by Radio Moscow. 2:18

13. Peru, panpipes and drum, collected by Casa de la Cultura, Lima. 0:52

14. "Melancholy Blues," performed by Louis Armstrong and his Hot Seven. 3:05

15. Azerbaijan S.S.R., bagpipes, recorded by Radio Moscow. 2:30

16. Stravinsky, *Rite of Spring*, "Sacrificial Dance," Columbia Symphony Orchestra, Igor Stravinsky, conductor. 4:35

17. Bach, *The Well-Tempered Clavier*, Book 2, Prelude and Fugue in C, no. 1, Glenn Gould, piano. 4:48

18. Beethoven, Fifth Symphony, first movement, the Philharmonia Orchestra, Otto Klemperer, conductor. 7:20
19. Bulgaria, "Izlel je Delyo Hagdutin," sung by Valya Balkanska. 4:59
20. Navajo Indians, "Night Chant," recorded by Willard Rhodes. 0:57
21. Holborne, *Pavans, Galliards, Almains, and Other Short Aeirs*, "The Fairie Round," performed by David Munrow and the Early Music Consort of London. 1:17
22. Solomon Islands, panpipes, collected by the Solomon Islands Broadcasting Service. 1:12
23. Peru, wedding song, recorded by John Cohen. 0:38
24. China, ch'in, "Flowing Streams," performed by Kuan P'ing-hu. 7:37
25. India, raga, "Jaat Kahan Ho," sung by Surshri Kesar Bai Kerkar. 3:30
26. "Dark Was the Night," written and performed by Blind Willie Johnson. 3:15
27. Beethoven, String Quartet no. 13 in B flat, op. 130, "Cavatina," performed by Budapest String Quartet. 6:37

THE NEXT LEVEL

The NASA *New Horizons* spacecraft, launched in 2006 and headed for a flyby past Pluto in July 2015, is also on an escape trajectory out of the solar system—the first such spacecraft on an escape trajectory since the *Voyagers*, and following a path similar to one of the Jupiter-Pluto missions that Gary Flandro and others charted in the mid-1960s. It is destined to continue on through a zone of thousands of small, icy planets beyond Neptune called the Kuiper Belt and enter interstellar space sometime in the next few decades. But it was launched without an interstellar message like *Voyager*'s on board. Perhaps this is a sign of a more anxious age.

In any case, a group of people led by Jon Lomberg are awaiting expected approval by NASA to upload a yet-to-be-determined "digital interstellar message" into the *New Horizons* spacecraft's permanent long-term flash memory once the mission has completed its science objectives. More than ten thousand people from 140 countries signed online petitions to support bringing this message project forward to NASA and the *New Horizons* project, which no doubt helped the idea gain official approval. The contents of the message—its text, images, art, and/or music—will be crowd-sourced, a distinctively more modern way of soliciting multiple opinions through the Internet. "Previous messages from Earth, portraits of our planet and our species, have been made by small groups of experts," Jon Lomberg noted. "This initiative proposes that this time, for the first time, the whole world can participate. The *Voyager* record has become an iconic image of the twentieth century, signifying our emergence as a galactic species. Now, new generations can be captivated by the incredible perspective that creating a self-portrait of Earth offers, becoming better informed citizens of the galaxy in the process."

I asked Jon to reflect on his motivation for taking advantage of the rare opportunity to once again include messages on an artifact being sent beyond our solar system. "Unfortunately, the *New Horizons* team was so busy just trying to keep their mission on track—it was canceled, then re-approved, then re-canceled, then re-approved—and successfully built and launched that they just didn't have the time needed to create a physical artifact like a plaque or a record," he told me. He was disappointed about that for a little while, assuming that it had to be some modern-day artifact equivalent to the Golden Record ("maybe a quantum nano superconducting *Voyager* record or something"). But then he thought more about it. *Voyager* had set the bar high for analog, physical artifacts. Later projects, like The Planetary Society's "Visions of Mars" on the NASA *Phoenix* lander, or the society's other efforts to launch the signatures of thousands of people on planetary missions, advanced that technology to CDs and DVDs. "But we have never sent out a digital message," Jon recalled thinking. "Nobody thought of putting one in the computer. So it's kind of the next level up from the physical artifacts. Granted, the lifetime of it probably isn't as long as the *Voyager* record, but again it's still an important gesture. Every spacecraft leaving Earth will have some type of computer, and so we may be establishing a positive precedent with the *New Horizons* digital message, especially the crowd-sourcing aspect of it. Before, it was just a few of us who were attempting to speak for the Earth. But with *New Horizons* we're making a serious effort to involve as much of the Earth as we can. And that's certainly something I think Carl would have liked."

I'm a member of Jon Lomberg's advisory board for what is being called the One Earth: New Horizons Message Project, and as we begin ramping up our public engagement in formulating what some are calling the *Voyager* Golden Record 2.0 message, it will be interesting to see how different today's crowd-sourced message to the future will be from the message so carefully crafted forty years ago by a select group of people for the *Voyager* record. "Perhaps *New Horizons* will never be found and its message never read," says Jon Lomberg, "but the very act of creating the message and sending it inspires the imagination and encourages a wider perspective in space and time. Humans have never needed this perspective more than they do today. Contemplating the vastness of the cosmos, we make our mark on it by our explorations—surely one of the most positive acts by the human species in all our history."

Part Two

THE GRAND TOUR

4

New Worlds
among the King's Court

NINETY-NINE POINT EIGHT percent of everything in our solar system is inside the sun, and of the 0.2 percent that is left, more than half of *that* is inside the planet Jupiter. Jupiter has more mass than all of the other planets, moons, comets, asteroids, and space dust out there combined, making its royal monikers from classical mythology— Zeus, Thor, King of the Planets—truly apt.

Jupiter was the first encounter for both *Voyagers*, and even after the successful *Pioneer* flybys just a few years earlier, much was still unknown and mysterious about our solar system's largest planet. In 1610, Galileo was the first to recognize Jupiter as a world with moons of its own, and the astronomers Robert Hooke and Giovanni Cassini were the first to recognize (separately), in 1665, the famous Great Red Spot and other colorful moving zones and belts of clouds in the giant planet's dynamic atmosphere. Over the intervening three centuries, improving telescopic resolution and instrumentation provided more information about the speed of Jupiter's winds and giant storm systems, of which the Great Red Spot is one, and about the chemistry of the clouds and the composition of the brightest moons. Right up until the *Voyager* flybys, however, those moons were still only points of light.

The *Voyagers* changed all that, forever. After the flybys, Jupiter's four large moons—Io, Europa, Ganymede, and Callisto, collectively known as the Galilean satellites after their original discoverer—became distinct worlds of their own, with features and characteristics and even personalities that now make many consider them full-fledged planets. Indeed, I believe that *Voyager*'s exploration of the Galilean satellites revealed a bias that we didn't even know we'd had in our search for life beyond Earth. The only example of life as we know it exists on a *planet*—a large body directly orbiting the sun—and a planet relatively close to the sun at that. Life on Earth takes advantage of our inner solar system location, of our abundance of liquid water, and the ample energy of sunlight bathing our planet. Therefore, why wouldn't we think it most likely that extraterrestrial life is planet-based as well? The only large moon in the inner solar system is our own moon, which lacks an atmosphere and thus we know is lifeless. But what if there are other moons out there that might have the right ingredients—liquid water, for example, or ways to tap into enough sunlight or other energy sources like volcanoes or tidal energy—to fuel the biochemistry of life? There are more than a dozen large moons in the outer solar system. According to the official definition, moons can be only once-removed cousins of planets because they revolve around a planet instead of around the sun. But what if that didn't matter? What if what matters instead in the search for life "out there" is what they are intrinsically like, not who they happen to hang around with?

FLIGHT PATH

Voyager mission designers such as Charley Kohlhase and his team of about ten colleagues at JPL had the job of figuring out how to time, align, and visualize the trajectories of each *Voyager*'s single pass through the Jupiter system so that the spacecraft would get the best possible views of the planet and its large moons, have a good communication geometry with the Earth, *as well as* have its trajectory bent and sped up by the right amount to swing the probe on to Saturn. Physicists realized long before Gary Flandro, Charley Kohlhase, and others on the *Voyager* team that such slingshots were possible, and that they represented the closest thing to a free lunch that one could get in the solar system. By aiming a spacecraft to pass *behind* a massive planet in its orbital path around the sun, the spacecraft would not only speed up as gravity draws the craft inward toward the planet, but it would also get a boost—a gravity assist—from the planet's own orbital momentum around the sun. It's kind of like the way a batter adds energy to a pitched softball when she hits it. The ball does not simply bounce off the bat with the same speed in the opposite direction—*energy* supplied by the batter is *added* to the ball's energy, changing its direction *and* increasing its speed. A planet's orbital momentum is a source of energy that a spacecraft can tap into to speed up (or, if passing *in front* of a planet relative to its direction of motion, to lose energy and *slow down* using the antigravity effect) relative to the sun. It seems like getting something for nothing, but it's not. Newton's laws of motion tell us that when it comes to forces and energies, there is always an equal and opposite reaction to any action. So for the spacecraft to speed up, it means that the planet has to slow down. Energy is conserved, never lost. The difference is, though, that because the mass of the spacecraft is so minuscule compared to the mass of the planet, the result of the spacecraft stealing some of the planet's orbital momentum (mass times velocity) is insignificant for the planet. When the *Voyagers* did their gravity-assist slingshots past Jupiter, for example, they were sped up by about 10 miles per second relative to their approach velocity, but Jupiter itself was slowed down by only the equivalent of about 1 foot per *trillion years*.

Setting up the flight trajectories for the *Voyagers* was a monumental task, sifting through what Charley describes as "10,000 possible flight paths" just for their primary mission targets Jupiter, Saturn, and Titan. Charley emphasizes that his team had to develop new software methods to quickly simulate and visualize many possible missions. One method included modeling spacecraft orbits using a centuries-old shortcut that broke the orbits down into shorter, simpler segments called conic sections ("Who was it, Kepler or Newton," Charley asked himself, thinking back on the early history of celestial mechanics, "who first came up with this mathematical trick?"—it was Newton who discovered the shortcut), because traditional orbit-calculation methods would have taken months to complete if using the full calculations and the computer technology at the time. "I've got a bunch of mission constraints I'm trying to honor, like communications, navigation, and getting the trajectory to the next planet," he recalls, adding wryly, "and I also know that the scientists would rather fly by the lit side of a planet and its moons rather than the dark side. And so we're trying to get the flybys close to these new worlds, as close as we can, but not so close as to magnify the navigation errors."

That last part was important: they knew that Jupiter would bend the trajectory of the spacecraft by a given amount because the mass of Jupiter was well known. But they didn't know the precise masses of the moons they wanted to get close to, and so they had to be careful not to get *too close*, lest that unknown mass bend *Voyager*'s path astray. "So we applied the engineering constraints, then what we thought would be attractive to Science." The team would then generate those cases in reams of plots and tables and pass them on to Ed Stone's Science Steering Committee to look over and give feedback as to the quality and scientific value of the various moon and planet geometries that these "computer-flown missions" would yield.

Charley positively beams with pride at his team's accomplishment of finding the two perfect needles in the haystack of mission designs they started out with: "Winnowing through that list of 10,000 possible missions to find the best 110 to target and the 2 to launch was an effort done nearly perfectly. I should show more modesty than that, but we did that job right."

The final March 5, 1979, Jupiter flyby path that Charley and Ed and colleagues on the Science Steering Committee chose for *Voyager 1* enabled close passes by Io, Ganymede, and Callisto, but only a relatively distant view of Europa. A more detailed view of Europa would have to wait until the July 9, 1979, flyby of *Voyager 2*, which made close passes by that moon, as well as Ganymede and Callisto, but allowed only distant views of Io. Thus it was only together, through both *Voyager* flybys, that high-resolution photos of all these worlds; movies and high-resolution photos of Jupiter's clouds and storms; and lots of other unprecedented data on radiation, magnetic fields, and the chemistry/composition of the Jovian system could be obtained.

Even at their super-high speeds of around 35,000 miles per hour, each of the *Voyagers* took about three days to pass through the heart of Jupiter's mini solar system, traveling from the orbital distance of farthest-away Callisto to their closest approaches to the giant planet (a distance away of only about three to five times the radius of the planet), then back out again. During that time, the spacecraft was in frequent communication with the DSN, radioing

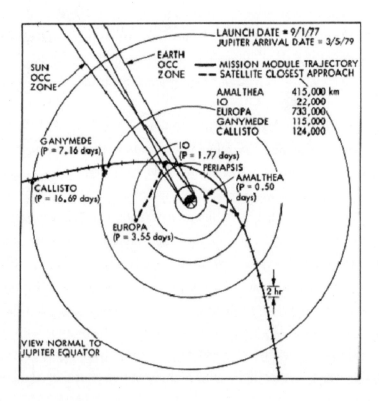

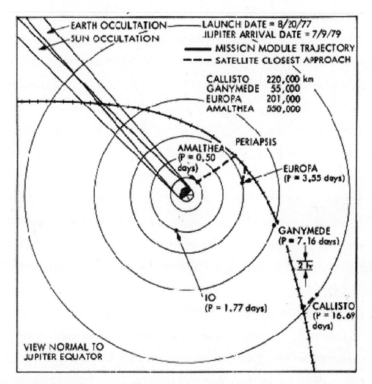

Jupiter Bound. *Voyager 1 (facing page)* and *Voyager 2 (above)* flyby trajectories past Jupiter. *(NASA/JPL)*

the latest images and other data back to an eagerly awaiting science team and press corps at JPL, and receiving updates to the onboard sequences with the newest team estimates of the best scan-platform pointing, camera exposure times, and other parameters.

All was going well as *Voyager* approached its first moon, Io. We saw weird splotches that weren't craters, but what were they? Io didn't look like we expected; what was down there? Everyone was excited. But then something went wrong.

Candy Hansen and her colleagues on the imaging team had learned from the *Pioneer* flybys of Jupiter that the radiation environment was harsh. Charley Kohlhase told me that the space-craft engineers had taken extra measures to shield the most sensitive electronics from radiation by using nearly fifty pounds of extra shielding made out of tantalum. Tantalum is a so-called transition metal because it can easily give up electrons to other atoms, thereby transitioning to different energy levels. It is not far from tungsten, molybdenum, and zirconium on the periodic table, making it a relatively dense metal. It is much more strongly resistant to corrosion than lead or gold and particularly good at blocking the kinds of high-energy solar proton and galactic cosmic-ray radiation that is commonly encountered in space missions. Still, despite this high-tech shielding, the intense bombardment of high-energy particles from Jupiter's enormous magnetic field was still having an effect on *Voyager*'s computers.

Specifically, the radiation was making the clock in the computer that controlled the scan platform get very slightly out of sync with the separate clock in the computer that controlled the camera's exposure time—telling the camera when to open and close the shutter. Computers work on the principle of binary numbers, ones and zeros, stored in voltages (electrons) inside transistors and microchips in a specific pattern that represents their software. Changing that pattern dramatically—by shutting off the power, for example—is essentially catastrophic to the software. But changing that pattern in very minor ways—for example, flipping a one to a zero here or a few zeros to a one there—may not be catastrophic but instead can make the software behave differently from how it's supposed to. This is what radiation (high-energy protons and atomic nuclei) can do: those particles can burrow down into a computer chip and strip off electrons, flipping a one to a zero, or cause new electrons to be added in, flipping a zero to a one. If this damage is localized to certain unlucky places in the software, the resulting change in the software could vary widely, from catastrophic and obvious to insidious and difficult to recognize. Software and hardware designers try to guard against the catastrophic by using radiation shielding, as well as multiple copies of the most critical software—the code that, if changed, could cause a catastrophe—in the computer. But it's much harder to guard against minor, insidious changes to the code, like a very small change in the rate of a clock in one computer as compared to another.

In some of the early *Voyager 1* images, particularly of Io, the out-of-sync computers caused the scan platform to start slewing to the next target while the camera's shutter was still open. The resulting images were hopelessly smeared, and the team "lost some really nice satellite data," laments Candy. "In addition to dealing with the anomaly it was stressful because there were so many 'guest' scientists in town and they were complaining bitterly, not understanding that we had not anticipated this situation." Even if they had understood—and once the explanation was found, they all did—I suspect that they were still upset and frustrated simply because of the loss of precious, one-time-only measurements as the spacecraft sped through the Jupiter system.

After years of working closely with the specific quirks and personalities of the *Voyager* spacecraft and system, and especially in the lead-up to the big events like the planetary flybys, Candy had begun to think of each of the spacecraft almost like a child. "You coach them and coach them to perform at the big school pageant," she says, "and then at the moment the performance is on you just sit in the audience, helpless, mouthing their lines and praying that they remember the tricky dance step. In that moment during the first Jupiter flyby, it was like the chandelier came crashing down in the middle of the performance." Rather than panic, however, she and her colleagues realized that they had time to devise a work-around. They had discovered the problem early enough in the encounter. They learned from this early experience, compensated for the out-of-sync computers, quickly uploaded new, last-minute versions of the camera sequences (time for Candy to bake more sorry-for-the-extra-work cookies!), and were able to avoid such problems for the rest of the mission.

With the Internet not yet invented, the only way for anyone not on the team to see all the images was to watch the "live" feed from the DSN on the monitors in the science, press, or public areas scattered around JPL or on a few of the team members' college campuses, such as at Caltech. Some of the coolest-looking photos would eventually get aired on the evening news or printed in the paper. Later, more would appear in magazines such as *Scientific American*, *Astronomy*, or *Sky & Telescope*, but only months after the images came in. The *Voyager* team needed to have near-instant access to the images, of course, to make sure the cameras were working properly, to make sure the spacecraft was on the right trajectory, and to try to do some "instant science" to share with the public and eagerly awaiting media at each day's press briefing. "Back in those days—it's amazing to think about now—we didn't have laptops or Photoshop," says Candy. "We had two small rooms, like broom closets, without windows, each with a display terminal hardwired to the image processing building where the telemetry came in and was turned into black-and-white images. We called those the browse rooms. You could call up the latest images and do simple contrast stretches.

The rooms were shared by every scientist on the imaging team." According to Candy, the browse rooms were huge bottlenecks for the imaging team, and they were occupied 24/7. "There was huge pressure to get press releases formulated, images processed, and captions written. The more senior members, of course, had higher priority, so all of us 'youngsters' would stay late into the night to have our turn." Still, she recalls with a wistful smile and twinkle in her eye, "That was actually a lot of fun."

Analyzing *every* image and other measurement as they were sent back over that critical three-day encounter period was the job of the members of the *Voyager* science and engineering teams. Some of the images were taken for purely engineering purposes, such as calibration or navigation. Many of the images that the *Voyagers* took when they were far from their planetary targets were taken through what's known as the Clear filter—a filter that let in the maximum amount of red, green, blue, and infrared light—so that faint stars could be photographed and the cameras used like a high-tech sextant to make sure the ship was sailing in the right direction. The *Voyagers* are what those in the business call a *3-axis stabilized* spacecraft. That is, they do not use a rotisserie-like spinning motion (*spin-stabilized*) like some other spacecraft to maintain their orientation. Instead, using thrusters, special sun sensors, and special star trackers, the spacecraft is held in a fixed orientation relative to the stars. Often, spacecraft will use the sun and one or two bright stars, such as Canopus or Sirius, to establish and hold a fixed orientation (or "attitude") while cruising through space. One nice advantage of this kind of attitude configuration is that cameras and other instruments can set very long exposure times for faint objects, without having to worry about the spacecraft's motion blurring the data.

However, sometimes the spacecraft needs to be pointed at an area of space with no bright stars. In that case, the science cameras can often be used to take photos of faint stars instead (there are always fainter and fainter stars to be found, anywhere in the sky). Whether imaging bright or dim stars for this purpose, this process is called *optical navigation*, or sometimes just *opnav*. Once that spacecraft-sun-star orientation is known—and it's the job of the navigation team to know it—spacecraft- or camera-operations people can point the equipment to any other known target in the sky with confidence.

Still, most of the photos taken were for scientific purposes rather than for celestial navigation, especially around the time of closest approach to *Voyager*'s various planetary and moon targets. Science images were often taken through color filters so that the red, green, and blue images could later be combined into a color image simulating what we would have seen with our own eyes if we were riding along on *Voyager*. Sometimes the team created "false color" images by using additional violet or infrared filters to produce more garish enhancements of what would have otherwise been only subtle colors to our eyes. In addition to their artistic or aesthetic value, color images also helped provide some information on the composition of the moons or the cloud layers in the atmospheres of the giant planets. Many times the colors or tones of features in the *Voyager* images required special image processing to bring them into view, however. This was often the job of staff scientists and engineers who worked in JPL's Image Processing Lab, where they had a browse room of their own.

Indeed, it was the astute pair of eyes of engineering and spacecraft navigation team member Linda Morabito that made the first, and perhaps one of the most important, of *Voyager 1*'s discoveries at

Jupiter's innermost moon, Io. It had been known since Galileo's time that the three largest innermost moons of Jupiter travel in a very special kind of orbital dance called a *resonance*. Specifically, for every single orbit of Ganymede around Jupiter, Europa orbits exactly twice and Io orbits exactly four times. That is, like the hour, minute, and second hands of a clock, those three worlds occasionally line up with one another, and thus their gravitational attractions nudge them each away from what would otherwise be perfectly circular orbits. Each of them, then, is sometimes slightly closer to or slightly farther from Jupiter than usual.

The mathematics of the orbital resonance of Io, Europa, and Ganymede had been worked out in detail around 1800 by the French astronomer Pierre-Simon Laplace (indeed, the resonance is named after him). But the implications of the Laplace resonance weren't fully appreciated until just before the *Voyagers* arrived at Jupiter. In fact, in a scientific publication intentionally timed to appear in print just three days before *Voyager 1*'s flyby, a team of three celestial mechanics experts led by Stan Peale of UC Santa Barbara published a prediction in *Science* magazine that the resonance that was slightly changing the inner Galilean satellites' distances from Jupiter would result in a gentle squeezing and relaxing of their interiors. Over time, the squeezing should heat the insides of those moons, with the strongest heating—perhaps all the way to the melting point—happening at Io, the moon closest-in to Jupiter. In one of the most famous modern examples of theorists making a testable prediction that eventually was dramatically proven to be correct, the authors ended their prescient paper by saying, "*Voyager* images of Io may reveal evidence for a planetary structure and history dramatically different from any previously observed."

And holy cow, were they right! Io was revealed to be like no planet ever seen before, or since. Measured to be only slightly larger than our moon, its surface is an alien reddish-brown in color and covered in lots of circular or semicircular splotches of yellow, white, and black. The splotches don't look like they are impact craters, however; in fact, *no* craters like those peppering the surfaces of our moon and Mars could be identified in the Io images. Because impact crater scars build up slowly and systematically on a planetary surface over time, the lack of any craters on Io suggests that its surface is constantly being refreshed, wiped clean by some process, and thus is very young. Io's mass could be estimated by measuring the tiny tweak that its gravity gave *Voyager 1* as it passed only 13,000 miles above the surface; combined with an estimate of its volume, this led to an estimate of Io's density as 3.5 grams per cubic centimeter. This number suggested a very rocky, rather than an expected icy, composition. Io was proving to be strange indeed!

The kicker, though, came from *Voyager 1* images taken three days after its closest approach to Jupiter, looking back toward Io. Linda Morabito had been tasked with helping to verify the post-flyby trajectory of *Voyager 1* using opnav images of the positions of faint stars seen in the backgrounds of images, in this case images of Io. What she found in one of those images, which no one had noticed previously, was stunning.

A little digression about *Voyager*'s digital images and the team's image-processing methods seems warranted here, to put the circumstances surrounding Morabito's discovery in context. The *Voyager* cameras took images at a resolution of 800 x 800 pixels, where each pixel could have a value between 0 (no signal) to 255 (maximum signal). According to *Voyager* imaging team member Torrence

Johnson of JPL, almost all the *Voyager* images that were being streamed to the science team on the TV monitors and printed out for team members in the science workroom were being displayed in black and white, where black meant signal levels near 0, white meant signal levels near 255, and various shades of gray corresponded to values in between. That is, they were being displayed using the full dynamic range of the cameras, taking advantage of the hard work done earlier by the science and sequencing teams to make sure they got the exposure levels right. These were great for normal pictures of Jupiter and Io and other moons, but they weren't very good for Morabito's search for background stars in the Io images. That was because the background stars were really dim compared to Io, maybe at signal levels near only 5 or 10, and so unless something different was done, they'd come out on-screen and in printouts looking basically black. Thus, in order to see the stars, either the navigation team had to request images with extra-long exposure times to make the stars brighter (and, perhaps, saturate the pixels of anything else in the scene, like a planet or moon) or the navigation team technicians would have to do what image-processing people call *stretching* the images, that is, changing the display so that black is still near zero but white is set to a much lower level, like 10 or 20—making the stars show up. Of course, everything else in an Io image with values above 10 or 20 would also show up as white, making Io itself look washed out. But that would be OK; it was the stars they were after.

When Linda Morabito viewed the Io images taken for navigation purposes, displaying and stretching them on the Image Processing Lab's workstation, the stars popped out as expected, but so did two other unexpected things: a bright circular blob along the day/night boundary on Io, and a fainter umbrella-shaped crescent sticking up a few hundred miles above the edge of Io's limb. The cloudlike feature above the limb looks remarkably like another moon passing behind Io, but Morabito and other *Voyager* navigation engineers knew that there weren't any other moons in the right place at the right time to explain that feature. There also weren't any camera smudges or other artifacts that would look like that. After ruling out those and other ideas, she and her nav team colleagues were left with only one hypothesis that seemed to fit the data: the crescent and bright blob were eruption plumes from active volcanoes on Io.

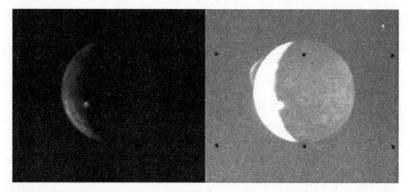

***Voyager 1* Io Volcano Discovery Image.** Image C1648109, revealing the first evidence ever found for extraterrestrial volcanism. Displayed at left as originally seen in the rolling displays broadcast on the *Voyager* science team monitors and at right in the harshly stretched format first used by the *Voyager* navigation team. The black dots are reseau marks embedded in the camera and used to correct slight image distortion. (*NASA/JPL/Jim Bell*)

JPL director Bruce Murray was skeptical, recalls teammate Torrence Johnson, as Murray had spent considerable effort dismissing similar claims of active volcanism on the Martian volcano Olympus Mons, claims based on fuzzy, cloudlike features in the *Mariner* mission images. Johnson recalled that imaging team lead Brad Smith tasked a select subset of the team, headed by Cornell planetary scientist Joe Veverka, to examine all the relevant Io images in greater detail, to try to confirm Morabito's hypothesis. Not only did Veverka's subteam confirm the volcanic plume nature of two features in that image, they quickly identified seven *additional* plumes in earlier *Voyager* images of Io, using similar image-stretching techniques as Morabito and the navigation team.

"The reason no one noticed the plumes earlier," Torrence Johnson says, "was because the real-time images we were seeing on the monitors during the encounter had been processed on the fly to cut off the top and bottom 5 percent of the pixels. Effectively, we had an *anti-volcanic plume filter* on the science monitors!" In each new plume discovery, the plume was located near or above dark surface depressions, in places where *Voyager*'s thermal infrared instruments were finding strange signals as well.

"I still remember the first Io data," recalls Linda Spilker, who was responsible for planning the infrared spectrometer observations. "The spectra had an unexpected slope because we were actually seeing both the Io background temperature and the much hotter volcano temperatures, but we didn't know it at first. We kept checking and rechecking the calibration until finally, with the discovery of the volcanoes in the images, we knew our data were right." The very high temperatures were consistent with molten or cooling lava. Supporting lab experiments showed that molten volcanic rocks containing large amounts of sulfur at different temperatures could reproduce the palette of white, yellow, red, orange, and black hues seen on Io's surface in the color images.

Io was a volcanically active world! Peale and his colleagues were right in their prediction published just days before the flyby—all that flexing from the combined effects of the satellite resonances and the strong tidal pull of Jupiter *did* heat up and melt the inside of Io. It was the first major discovery of the mission, and the first discovery of active volcanoes beyond the Earth. Four months later, when *Voyager 2* flew through the Jupiter system and photographed Io, the surface had changed significantly, including the formation of new plumes. Images taken by three more spacecraft, including the *Galileo* Jupiter orbiter, that have studied Io in the decades since have shown even more changes in the moon's tortured volcanic surface. Not only is Io volcanically active, it is *hyperactive*, harboring the most intense and voluminous volcanic eruptions in the solar system.

"We're talking about a moon whose geology changes the same way the weather changes on our planet," points out Rich Terrile. "It's something right out of science fiction. And yet, it's right here, orbiting Jupiter." The little moon is turning itself inside out trying to get rid of all that internal heat.

My JPL planetary science colleague Rosaly Lopes recalls, "I was a student at the time *Voyager* discovered Io's volcanoes, and I thought that an incredible discovery. *Voyager* detected a dozen active volcanoes, and that blew our minds at the time." Rosaly later earned a spot in the 2006 edition of *The Guinness Book of World Records* as the discoverer of the most active volcanoes anywhere—a total of seventy-one on Io.

Voyager's images at Jupiter's second-closest big moon, Europa, were also full of surprises, although what was found to lie *under* the surface is what gained the most attention, then and since. Europa was only rather poorly photographed by the *Pioneer* missions earlier in the 1970s, and *Voyager 1* was able to see its bright icy surface only from a long distance away since its trajectory was optimized for close passes by Io, Ganymede, and Callisto. Four months later, *Voyager 2* passed through the Jupiter system and got close-up views of Europa. Oh my goodness, was it worth the wait! The details seen in *Voyager 2*'s photos were truly new and exciting. Indeed, perhaps the most common first impression among the imaging team when seeing those first close-up photos was "Wow—that's *flat!*" And it sure is. Europa is about the same size as our moon, but unlike the 3 to 5 miles of rugged elevation difference found among the mountains and valleys of our moon, the largest "mountains" and deepest "canyons" on Europa are only around 30 to 50 feet tall or deep. That is to say, if Europa were the size of a bowling ball, the tallest bump on its surface would be less than the thickness of a piece of thread! Another surprise was (again) the relative lack of impact craters—the scars left on ancient planetary surfaces by asteroid and comet impacts over the eons. This implied that some process must be resurfacing Europa, erasing the craters that must surely have built up over time. To everyone's amazement, *Voyager* had discovered one of the flattest and youngest surfaces (though not as young as nearby Io) in the solar system.

But why is it so flat? A clue came from the crazy-quilt series of lines that crisscross Europa's surface—dark cracks that separate the icy crust into a jumble of curvy or triangular sections, almost like the pieces of a big jigsaw puzzle. In some places, the puzzle pieces seem to have rotated relative to one another, and in other places to have spread apart, letting reddish-brown material ooze up in the intervening spaces, maybe similar to the way fresh, new volcanic lava oozes up between tectonic plates that make up the Earth's mid-ocean ridges. Indeed, one of the most common and obvious reactions to Europa's cracked and platelike surface was to compare it to melting sea ice—a thin layer of frozen water floating on top of liquid water, sloshing around. On Earth, waves and the warmth of summer break up polar sea ice into millions of little icy "plates." If that is essentially what is happening on Europa as well, the implications would be enormous. Life on Earth may have begun in the ocean. Does Europa have one?

Frustratingly, the *Voyager* view of Europa was only fleeting—two quick flybys, and one at a pretty far distance away. The known presence of strong heating from tidal forces (heavily in evidence at Io), the super-flat surface, the sea ice–like plates with distinctively colored material appearing to ooze up from the depths below—all these pieces of evidence pointed toward the possibility of a subsurface ocean. I like to imagine what would have happened if *Voyager* had been outfitted with a high-speed submarine probe that could have penetrated Europa's thin ice shell and plunged into the watery depths below: Turning on its headlights, the sub relays real-time video as it dives deeper and deeper. Finally approaching the seafloor, the water becomes murky, and the probe's thermal sensor detects a hot spot up ahead. Jupiter's tidal energy is flexing Europa, heating its icy and rocky interior, and the chemical sensor identifies hot, sulfur-rich water and gases leaking out of the crust here just like at many mid-ocean ridges on Earth. With the outside pressure rising and beginning to stress the sub's hull, the onboard mass spectrometer starts an analysis for organic compounds in the hot waters. Switching to wide-angle mode, the probe begins to scan for any signs of motion. The pressure is getting critically high and the sub's signal is beginning to weaken as we're now more than 60 miles below the surface. The video is getting ratty. There's a flash of some kind—then static—then another flash, and then a strange, curved silhouette of—something? But then the signal is lost and we all just stare, dumbstruck, at the static. What had we just seen?

"Launch a class-five probe, Number One!" *Star Trek*'s Captain Picard would have ordered, had the USS *Enterprise* encountered this ice-covered water world. "And fit it for higher-pressure submarine operations!"

But as Europa receded from view, all the *Voyager* team could do was make the best of the limited data in hand and dream about the day when they'd be able to go back, at least virtually, and take a longer, deeper look at this enigmatic world. US Geological Survey planetary geologist and *Voyager* imaging team satellites subteam lead Larry Soderblom remembers the sort of deer-in-the-headlights feeling that he and many others on the imaging team had after taking only those two brief passes by Europa. Although he is an expert geologist, his Earthbound experience left him stumped time and again when trying to interpret the strange new landscapes that were revealed in the *Voyager* images.

"Although we only had a glimpse of Europa, the fact that there were so few impact craters on its surface left no doubt that its surface was geologically young—maybe only 100 million years old," he explains, being sure to note that 100 million years truly is "young" to geologists. "Europa formed about four and a half billion years ago along with the rest of the solar system, so 100 million years is only about 2 percent of its lifetime. Surely, we all thought, its surface must still be changing today. But what causes those changes? We all wished we'd had a chance to take a closer look."

Imagine for a moment that you had been dreaming all your life of someday seeing the Grand Canyon in person. Then imagine you decided to walk there. You set foot to the pavement day after day, and after half a year, give or take, you made it! You would hike to the bottom, set down your tent, raft the Colorado River, and explore. But wait: what if your only option was to keep walking right by it, peering over the rim, and dreaming wistfully of seeing those spectacular layers of colored rocks and feeling that ice-cold water running through your toes? That's what Larry and others felt like after *Voyager*'s too-quick glimpses of Europa . . . we were so close and yet so far.

The chance for a closer look would not come for more than sixteen years, when NASA's *Galileo* Jupiter orbiter mission began making close flybys of Europa and the other Galilean satellites in 1995. *Galileo* had the advantage of spending *lots* of time in the Jupiter system, orbiting the giant planet thirty-five times, on trajectories that took the spacecraft close to Europa eleven times. High-resolution color images of Europa's cracks and other features, measurements of the variety of ices and minerals on the surface, and the discovery that the subsurface is electrically conductive all provided additional evidence for *Voyager*'s initial hypothesis of a deep, salty ocean under Europa's icy crust. The conductivity measurement in particular is especially intriguing, because just a few percent of dissolved salts (much like NaCl, or table salt) in a deep liquid water ocean could explain the measurement. That kind of salinity would make Europa's ocean very Earthlike. In fact, *Galileo* data are consistent with Europa having the largest ocean in the solar system—with maybe two or three times the volume of water in all the Earth's oceans combined.

Still, though, the evidence is indirect, and many of us yearn for *proof* and *details* about Europa's putative liquid-water ocean. Is it really there, or is there just more slushy ice under Europa's frozen crust? If it's really there, how deep is it? How warm is it at the bottom, where the tidally heated rocky part of Europa's crust is in contact with the water? Are there organic molecules in that ocean? Heat sources, organic molecules, liquid water: these are all the hallmarks of a *habitable* environment for life as we know it. Exciting, for sure, but evidence that a place is habitable is not necessarily evidence that it is *inhabited.* Is there life in Europa's ocean?

I feel the same way the *Voyager* team did in 1979: we have to go back! And we have to go back for a longer visit, a dedicated visit, to find out. We can send missions to do more flybys and eventually to orbit Europa and map its surface in detail, to map the thickness of the icy shell of frozen water and find the places where it's thinnest. We can land a mission there, perhaps robotic, perhaps human-crewed, and drill into that thin ice and find proof of the ocean below. If it's there, and if we can get through the overlying ice, we can send a real version of my imaginary submarine down there and take pictures and make chemical and biologic measurements, maybe even collect samples to bring back to Earth. Oh, for one of Captain Picard's class-five probes! In the decades ahead, we are in for a grand adventure exploring the number-one nearby locale in the search for living organisms beyond Earth. I predict that these missions will give us the answer about life in Europa's ocean. I am trying to eat well and exercise regularly so I can live to see that exploration pay off.

After the bizarre and unexpected revelations at Io and Europa, many on the *Voyager* team could easily have figured, That was it. How could it get any better? They had discovered amazing secrets of the Jovian system. But still, *Voyager* marched on. Every precious moment of each spacecraft's three-day plunge past the giant planet and its moons had been filled with the maximum number of photos and other measurements that the power supply and tape recorder could handle. It would have been an incredible set of sights to behold if we could have magically traveled aboard *Voyager* and looked over the shoulder of those cameras as they snapped their timeless photos, revealing strange and lovely new vistas at every turn of the scan platform.

Passing by Ganymede, the largest moon in the solar system (larger than the planet Mercury!), both *Voyagers* revealed evidence for past movement of a grooved, icy, platelike crust similar in some ways to Europa's but apparently much more ancient because the many impact craters on its surface had been preserved in the ice. Ganymede was apparently not subject to the constant oozing of material from below like Europa, leaving much of its cratered surface intact. The team speculated about the possibility of a subsurface ocean on Ganymede because it was also tidally heated by the orbital resonance along with Europa and Io, but no convincing evidence was seen in the flyby data. Instead, as it had for Europa, it took the more detailed and frequent flybys by the *Galileo* mission in the 1990s to discover that Ganymede has a magnetic field (the only moon in the solar system with its own) as well as a conductive subsurface layer under its icy surface—perhaps another salty ocean waiting to be confirmed. We'll have to wait a while to find out, however, as the next robotic mission to Ganymede, the European Space Agency's *Jupiter Icy Moons Explorer* or *JUICE* spacecraft, won't launch until 2022 and won't orbit Ganymede until 2030.

The one large Galilean moon that is not part of the resonant dance around Jupiter with the others is Callisto; while its surface is not as exciting as that of its large siblings, *Voyager* data nonetheless revealed that Callisto has mysteries of its own. Callisto's surface is heavily cratered, covered from stem to stern with impact scars from billions of years of pummeling by asteroids and comets. That observation alone helps us appreciate the significance of the nearly craterless surfaces of Io and Europa, and the mild cratering of the icy surface of Ganymede. Callisto, living as it does roughly in the same vicinity as those three other moons, tells us that its siblings had to have been smashed countless times as well. But their surfaces are so much younger and more dynamic that much or all of the evidence for those past impacts has been covered over or wiped away. Callisto's relative lack of internal heating has made it a more passive world, taking its blows in stride. One of those strikes photographed by *Voyager* is an enormous, more than 2,300-mile-wide multiringed basin called Valhalla that preserves evidence for a whopper of a giant impact early in Callisto's history. Despite its apparent lack of interesting surface geology, a more detailed study of Callisto by the later *Galileo* mission revealed evidence that there might even be a thin liquid water layer—an ocean of sorts—beneath that moon's thick icy crust.

Voyager and subsequent missions have shown us that Io, Europa, Ganymede, and Callisto are a sort of mini solar system revolving around their "sun," the giant planet Jupiter. Tidal forces from Jupiter and from one another, and perhaps some radioactivity from the moons' deep rocky and metallic cores, heat the insides of these worlds and lead to massive high-temperature eruptions of sulfur-bearing volcanic rock on Io and probably to liquid-water layers—subsurface oceans—on Europa and Ganymede and perhaps even Callisto. *Voyager*'s discoveries at Jupiter included other moons as well, including the first close-up views of the small potato-shaped moon Amalthea, the fifth known moon of Jupiter (discovered in 1892), and also the discovery of three new moons (Metis, Adrastea, and Thebe), all of which are too small and faint to be seen from Earth, and all of which also orbit close-in to the planet like Amalthea.

Maybe the most amazing "small moon" discovery, however, took advantage of *Voyager*'s unique perspective of being able to look back, sunward, toward Jupiter after having flown past on its closest approach. Anyone who's ever driven westward around sunset knows that driving into the sun's glare causes all the dust and grime and bugs on your windshield to light up, making it hard to see. This effect is known as *forward scattering*, and it was a trick that was exploited by the *Voyager* imaging team to try to search for small particles from dust while pointing the camera back toward the general direction of the sun. Lo and behold, the strategy worked, and a newly discovered set of thin, dark rings around Jupiter was spotted in *Voyager*'s images. They turned out to be very small "moons" indeed; most are just motes, and the largest ring particles are only about the width of the thinnest human hair.

Voyager's moon and ring discoveries were exciting and historic to be sure, but the clear highlight of the flybys in terms of sheer photographic beauty was the imaging of the planet itself. My former Caltech professor and research mentor Andy Ingersoll was one of the visionaries who helped plan a time-lapse movie of Jupiter's swirling storm clouds as the *Voyagers* approached the planet. From far away, *Voyager*'s pictures weren't much better than the best pictures that could be taken from Earth telescopes at the time. But as the cameras got closer, richer and subtler details began to emerge. The Great Red Spot, first seen more than three hundred years earlier, wasn't just a single storm but was instead revealed to consist of lots of smaller storms swirling within and around the edges of a giant, multilayered, multihued, high-pressure vortex more than three Earths across. Watching Andy's *Voyager* approach movies makes it feel as if you are riding along with the spacecraft, watching in wonder and perhaps a little fear as the imposing cyclones loom larger and larger. . . .

Voyager's closest-approach imaging of Jupiter's clouds provided more visual delights. Waves, swirls, spirals, and streaks waft across the photos like the mad flicks of Van Gogh's brush, painting a cosmic canvas scene like none before. Clouds that *Voyager* scientists later found to be made of ammonia, methane, hydrogen sulfide, phosphine, and even plain old water vapor dance upon a palette of reds, browns, yellows, and whites—swirling gracefully but at speeds of more than 200 miles per hour. The pressure and the turbulence would certainly shake to pieces any modern jetliner trying to cruise above those storms. It's a landscape that elicits awe and wonder.

Since the *Voyagers* flew past in 1979, three more robotic space probes have visited Jupiter. The *Galileo* orbiter arrived in 1995 and, despite a stuck radio antenna that severely limited the amount of data that could be sent back to Earth, it successfully explored the system until it nearly ran out of maneuvering fuel and was commanded to plunge into the clouds of Jupiter in 2003 (in NASA parlance, the spacecraft was "disposed of" within Jupiter, to avoid an accidental crash with and possible contamination of the potentially life-bearing ocean on Europa). As it descended deeper into the giant planet's endless crushing pressures, *Galileo* was eventually completely vaporized (atoms and molecules from our planet, and our handiwork, are now freely floating through those beautifully colored, gracefully swirling clouds). Since then, the *Cassini* mission flew past Jupiter in late 2000–early 2001, getting a gravity-assist kick on its way to Saturn; and the *New Horizons* mission flew past Jupiter in 2007, also getting a gravity-assist kick from the giant planet, helping to propel that spacecraft to higher speeds for a quicker flight time to Pluto, which it will fly past in 2015. Both the *Cassini* and *New Horizons* flyby missions took photos and other measurements of Jupiter and its moons, in a sort of modern redo of the *Voyager* flybys, but with more high-tech instruments and data-storage capability. More recently, NASA's *Juno* mission was launched in 2011 and is en route to a 2016 encounter with Jupiter. Once there, it will spend an Earth year orbiting the giant planet and studying its magnetic field, radiation environment, and gravity, building on the earlier *Voyager* and *Galileo* observations by providing new clues about the planet's deep interior and core.

Even very recently, there's been new excitement in the Jovian system: plumes of water vapor have been discovered emerging from the south pole of Europa. Over the past few years, astronomers have been using the Hubble Space Telescope to make a sensitive search (much more sensitive than had been possible with *Voyager*'s technology and trajectory) for water vapor near Europa. Several "puffs" of water vapor were seen in the Hubble data, coming and going in time with the gentle tidal stretching (puffs seen) and contracting (no puffs seen) of Europa's cracked surface ice.

"By far the simplest explanation for this water vapor is that it erupted from plumes on the surface of Europa," planetary astronomer and Hubble study lead Lorenz Roth wrote in the official NASA press release. Hubble data appear to show that Europa has plumes or jets coming out of at least some of the long, dark cracks in its icy crust, perhaps like those detected by the *Cassini* spacecraft around

Saturn's icy moon Enceladus. Roth went on to speculate that "if those plumes are connected with the subsurface water ocean we are confident exists under Europa's crust, then this means that future investigations can directly investigate the chemical makeup of Europa's potentially habitable environment without drilling through layers of ice. And that is tremendously exciting."

The *Voyager* flybys of Jupiter in March and July of 1979 provided a massive shift in our understanding of giant planets and their moons. Once thought to be too cold, too far from the sun, to support life as we know it, the Galilean satellites were found to have significant amounts of internal heating, fueled by tidal forces. Europa, Ganymede, and possibly even Callisto appear to have vast reservoirs of subsurface liquid water—oceans—under relatively thin shells of ice. Astrobiologists—scientists who study the origin, evolution, and fate of life on the Earth and potentially other planets—now think about Europa as one of the leading candidates on the short list of worlds beyond Earth where extraterrestrial life may exist, or may have existed long ago (other places on that short list include Mars, and Saturn's moons Titan and Enceladus).

Indeed, just in the last few years, support for a NASA mission back to the Jupiter system that would focus on Europa (and that would complement the Europeans' *JUICE* mission to Ganymede) has grown substantially. This support comes partly from the scientific community, which ranked a return mission to Europa as an extremely high priority for NASA in the most recent National Academies of Sciences "Decadal Survey of Planetary Science"; partly from extraordinary public and media interest in astrobiology and the search for life elsewhere in the universe; and partly (and unexpectedly) from the individual, personal interest of US congressman John Culberson, who represents Texas's Seventh District in the suburbs west of Houston. For reasons that I can describe only as remarkable and supremely fortunate for my line of work, Mr. Culberson just loves Europa. He is fascinated by the possibility of life on this ocean world. I've visited him in his office in Washington (he is perhaps the only person out of 535 members of Congress with photos of Europa on his office walls!) and, along with such *Voyager* team members as Candy Hansen (who recently served a term as the chair of the American Astronomical Society's Division for Planetary Sciences), have helped keep him updated on the latest findings about Europa from NASA's missions and telescopes. He is an educated and engaged advocate for space exploration—a rare and delightful occurrence in Congress, to be sure. But perhaps most important, he is also a high-ranking member of the US House of Representatives Committee on Science, and so he takes it as a personal goal and passion to try to convince his colleagues in Congress to authorize funds for NASA to go back to Europa and find out what it's really like. Mr. Culberson has been successful in the past few years in getting some funds allocated for the development of new technologies needed to operate spacecraft for long periods of time in Jupiter's high-radiation environment. Sometimes, even in rocket science, a personal touch makes all the difference.

5

Drama within the Rings

EVEN WITH TWO spectacular flybys of Jupiter in the bag, and a
trove of planetary science discoveries and puzzles opened in the
process, the *Voyager* team didn't feel like they had much time to rest
and reflect on their good fortune. Both probes were speeding on to
encounters with the giant ringed planet Saturn, with *Voyager 1*'s Saturn flyby set to occur only sixteen months after *Voyager 2*'s at Jupiter. That might seem like a lot of time, but every working hour was
spent on the planning that needed to be done to optimize the trajectories of the spacecraft past the planet's moons and rings.

I remember the first time I saw Saturn through a real astronomical telescope. I must have been around ten years old, and I think it
was during a Boy Scout field trip to a small rural observatory run by
a local amateur astronomy club (I think it was the SkyScrapers, the

club I would later join). It was a clunky old refractor—the kind of long telescope that uses lenses instead of the more modern shorter kind that uses mirrors—with a tube like an iron water main and giant counterweights and rivets that could have come right out of a 1930s WPA construction project. Still, it was a *real* telescope, capable of large magnification and good image quality, especially from a rural site on a clear summer night. The telescope operator was hopping around the sky, manually slewing the tube to focus on many of the greatest hits (which he knew by heart) for our viewing pleasure. Double stars, nebulae, star clusters, a faint, fuzzy comet ... Each of us had a few seconds of viewing time, perched up on a step stool so we could reach the eyepiece. When it was my turn to see Saturn, highly magnified, in steady skies, with rings tilted gloriously in view, I remember going into an almost trancelike state, hypnotized by the elegance of what I was seeing. The next kid in line started pushing me to step away so he could have his turn, which I did only reluctantly.

What I didn't realize at the time was that professional astronomers in the 1970s weren't really *that* far ahead of the amateurs in terms of knowing what Saturn was really like. It's funny how we always think that the experts are leagues beyond the rest of us, members of a secret club that we can never enter. Knowledge of the chemistry of Saturn's atmosphere at the time was based on some good measurements from big telescopes, some assumed similarities with the chemistry of Jupiter, and the assumption that this gas giant, too, was likely mostly made of hydrogen and helium, like the sun (after all, it formed from the same cloud of gas and dust that formed the sun itself). Even less was known about the planet's famous rings and its collection of ten then-known moons. It was understood that the rings could not be a solid ring of material (the British physicist James Clerk Maxwell showed in the 1850s that such a ring would break apart from the stresses of the inner and outer parts orbiting at different speeds). It was also known that some of the moons were quite large, perhaps even planet-sized bodies. The first hints of the icy composition of these worlds were only just starting to come in from the latest high-tech spectrometers mounted on large Earth-based telescopes.

While the *Pioneer 11* flyby of the Saturn system in September 1979 helped us understand the gravity and magnetic fields of Saturn in much more detail, even that flyby encounter, with its limited imaging capabilities, didn't dramatically increase our knowledge of what Saturn and its rings and moons looked like in detail. Besides showing that Saturn's largest moon, Titan, is an extremely cold world—at only maybe 100 degrees or so above absolute zero, probably too cold for life as we know it—a key facet of *Pioneer 11*'s trip through the Saturn system was to simply demonstrate that it could be done.

"*Pioneer 11* was targeted to fly through Saturn's ring plane near where *Voyager 2* had to fly through the ring plane to go on to Uranus," says Ed Stone. *Pioneer* was only about a year or so ahead of the *Voyagers*, and so Ed and Charley Kohlhase and the rest of the *Voyager* team were paying careful attention to *Pioneer*'s mission, which was run by colleagues from NASA's Ames Research Center, just south of San Francisco. Among the questions scientists had for *Pioneer*: Could a spacecraft pass unscathed through the plane of Saturn's rings, not that far from the dense rings themselves? Would there be unanticipated radiation or magnetic field effects close to Saturn that could be more dangerous than what was encountered at Jupiter?

From below and beyond the rings, *Pioneer 11*'s farewell pictures of Saturn foretold of the even more spectacular sights and perspectives to come from the *Voyagers* that would follow. *Pioneer* had passed through the ring plane of Saturn without incident—but perhaps only just barely. Later analysis of the *Pioneer 11* trajectory showed that the spacecraft may have *just* missed (by about 2,500 miles—a near miss in astronomy) crashing into a small moon only later discovered orbiting near Saturn's rings. The experience caused some concern for *Voyager 2*, but mission planners were not particularly concerned.

"We thought, So it passed within 2,500 miles of something. So what?" Charley Kohlhase says. "Space is big!"

Regardless, because of the importance of the Titan flyby,

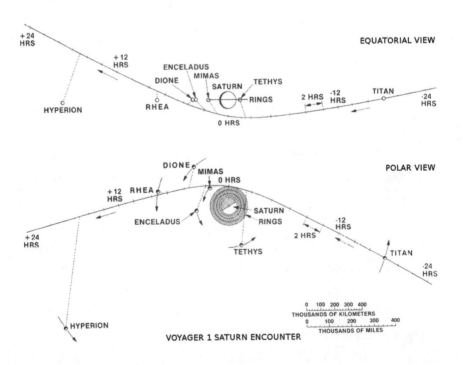

VOYAGER 1 SATURN ENCOUNTER

POLAR VIEW

EARTH OCCULTATION
ZONE

SUN OCCULTATION
ZONE

RING-PLANE
CROSSING

TETHYS

ENCELADUS

SATURN
CLOSEST
APPROACH

•MIMAS

•DIONE

RHEA

VOYAGER 2
SATURN
ENCOUNTER

200.000 km

2 HRS

TITAN

HYPERION

EARTH | SUN

Saturn Swingbys. *Voyager 1* (*facing page*) and *Voyager 2* (*above*) flyby trajectories past Saturn. (*NASA/JPL*)

Voyager 1 had to be targeted to cross the plane of Saturn's rings much farther from the planet itself than *Pioneer* had. *Voyager 2*, however, would have to take a deeper plunge through the ring plane— closer-in to the planet—if it was going to use Saturn's gravity to slingshot ahead to Uranus and Neptune. Just like at Jupiter, *Voyager* mission planners tried to time each spacecraft's two-day trip through the heart of the Saturn system to get as close to the planet, rings, and as many moons as possible, albeit with some important constraints.

An important constraint was imposed by Ed Stone and the *Voyager* science team: the trajectory had to take the spacecraft *behind* and *into the shadow of* the planet as seen from both the Earth and the sun, so that both sunlight traveling to the spacecraft, and the spacecraft's radio signal traveling to Earth, would pass through Saturn's upper atmosphere. Such an event is called an *occultation*, because the planet blocks or *occults* (obscures) the sun (or the Earth) from the viewpoint of the observer, which in this case is the *Voyager* spacecraft. An eclipse is a kind of occultation. Occultations of sunlight passing through a planetary atmosphere (which *Voyager* could observe) or *Voyager*'s radio signals passing through a planetary atmosphere (which the DSN could observe from Earth) provide a way for scientists to probe the details of the pressure, temperature, and chemistry of gases in that planet's atmosphere. As sunlight passes through the upper atmosphere, it is absorbed by gases deeper and deeper in the atmosphere until it is blocked entirely by the increasing density of the gas (for a giant planet), or by the surface itself (for a moon or terrestrial planet). Instruments on *Voyager* could measure the patterns of that absorbed sunlight and use those patterns as fingerprinting tools for the identification of specific atoms and molecules. The same is true of the DSN antennas watching the pattern of *Voyager*'s radio signal slowly fade from view as it went behind the planet from the Earth's perspective. It's a powerful scientific trick to exploit and so it couldn't be passed up when designing the optimum orbit trajectory. I can imagine that it was just another source of headaches sometimes for Charley Kohlhase and his mission-design team, however.

Even more important, perhaps, was the flyby of the moon Titan. What little information there was about Titan up until that point seemed to suggest that its environment could be similar to that of the early Earth. The flyby was a possible way to get in touch with our primordial past! That called for close-up imaging and other measurements of Titan, including a pass behind the moon as seen from the Earth and the sun. This requirement would almost single-handedly define the eventual trajectory and fate of *Voyager 1* past Saturn. Moreover, the possibility of *Voyager 2* continuing on to Uranus and Neptune would depend entirely on whether *Voyager 1*'s Titan encounter was successful.

Prior to the space age, Titan had been discovered to have an atmosphere (the only moon known to have one), consisting of at least methane and perhaps some other complex hydrocarbons. *Pioneer 11*'s low-resolution flyby images showed just an orangey sphere, bland but suggestive that the atmosphere was likely thick and hazy. Despite having extremely low temperatures, the evidence that Titan was a model of what the early Earth's atmosphere may have been like was significant. Before life began adding oxygen to our planet's atmosphere, Earth's atmosphere was also rich in hydrocarbon (and nitrogen), what chemists call a *reducing environment* (as opposed to an *oxidizing environment*).

In some pioneering experiments in the 1950s, biochemists Stanley Miller and Harold Urey did a famous set of experiments to demonstrate how adding water and energy (like lightning) to a reducing environment containing simple hydrocarbon gases could lead to the formation of even more complex organic molecules, including some simple amino acids. Biologists, and now astrobiologists as well, believe that this kind of chemistry could have led to the formation of life on Earth. *Voyager* scientists wondered if it could have led to similar kinds of biogeochemical magic on Titan. Maybe the oldest single-celled organisms in the solar system were swimming around in Titan's primordial soup.

So *Voyager* simply *had* to encounter this special, one-of-a-kind, Mercury-sized moon called Titan, up close, to measure its atmospheric pressure, temperature, and chemistry in more detail and to try to glimpse its frigid surface. Indeed, based partly on the *Pioneer 11* results from a year earlier, *Voyager 1* was targeted for a very close pass just 4,000 miles—or a little over one Titan diameter—above the haze. To help increase the odds of a successful flyby, Charley Kohlhase and the other mission designers also timed the Titan flyby to occur before *Voyager 1* passed through the plane of Saturn's rings and its closest approach to the planet. This "Titan before" approach also had the benefit of bending *Voyager 1*'s trajectory away from a potentially more risky dive deeper through the heart of Saturn's rings while still allowing encounters with many of the planet's other large moons. It was a hopefully safe course through poorly charted territory, focused on the prize: Titan.

The flyby went well. Titan's size and mass were measured directly, leading to an estimate of its density that implies that Titan is a world of rock and ice, like Europa and Ganymede, rather than ice alone. Atmospheric measurements found that Titan's air is indeed highly reducing—mostly made up of nitrogen, but also with significant amounts of hydrocarbon gases like methane, ethane, acetylene, and ethylene, as well as hydrogen cyanide. Many of these same kinds of gases may have dominated the early reducing atmosphere of our own planet. The surface temperature was found to be only around 90 degrees above absolute zero, but when combined with the high surface pressure found by *Voyager 1*—about 50 percent thicker than Earth's atmosphere—this led to one of the most surprising discoveries of the encounter: at those pressures and temperatures, methane, ethane, and many other hydrocarbons found on Titan can be gaseous, liquid, or solid. The ubiquitous haze covering Titan is thus likely due to thick, exotic *clouds* made out of methane and ethane!

And this is what led to the one and only disappointing part of *Voyager 1*'s encounter with the solar system's second-largest moon. That thick haze completely obscured the surface from view, covering up what many scientists believed could be spectacular vistas of rivers, lakes, even oceans of liquid hydrocarbons on the surface below. River valleys carved by liquid ethane! Waterfalls of methane! What sights were hidden from view by that blasted haze?! Without the capability to look through the clouds, for example by using radar like the meteorologists do on the evening weather report, *Voyager*'s cameras were blind to the surface itself.

If the spacecraft didn't have to aim for such a close Titan encounter, *Voyager 1* could have used a gravity assist to continue to travel on to Uranus and Neptune (as *Voyager 2* did). *Voyager 1* gave up a lot for that close flyby, and the fact is that *Voyager 2* might have skipped Uranus and Neptune if anyone had thought it could get a better look at Titan. That is how much emphasis had been put on finding out what Titan was really like.

I asked mission architect Charley Kohlhase what he thought of the official *Voyager* Project policy of giving up on *Voyager 2*'s Grand Tour if the *Voyager 1* Titan flyby had not been successful. He replied in a millisecond: "I hated it." There was a lot of unofficial support for the Grand Tour option (which helped Charley and the other mission designers keep that possibility on the table as they were developing their *Voyager 2* Jupiter and Saturn flyby scenarios), but also a lot of scientific interest in Titan as a potential model for the early Earth. Both Charley and I agreed that it's impossible to know which way that decision would have gone.

Ed Stone is more certain: "If *Voyager 1* had not worked, *Voyager 2* would have gone the same way, as a Jupiter-Saturn-Titan mission." Interestingly, an opportunity had been identified relatively early in *Voyager*'s mission planning to use the Saturn swingby to propel *Voyager 1* to a later encounter with Pluto—at the time the solar system's most distant known planet. However, the need for the close pass by Titan took that option off the table. Fortunately, Pluto still garners significant public and scientific interest, and so even though it has since been officially demoted from planethood, it was still judged worthy of a flyby mission of its own, and so the *New Frontiers* probe will give us, finally, a first glimpse of that distant former planet in the summer of 2015.

Horizon

ICE VOLCANOES

Titan was just the first step in *Voyager 1*'s path of discovery through the Saturn system. The team planned and acquired the first high-resolution images and other measurements of all of Saturn's other large moons—Tethys, Mimas, Enceladus, Dione, Rhea, and Iapetus—as well as the first detailed images of Saturn's famous rings, viewed on approach from above, then from edge-on, and then from below and behind. Before *Voyager*, the rings were thought to consist of just three major sections (imaginatively dubbed A, B, and C, and so on, from outermost to innermost), with empty gaps between them. But all that changed after the *Voyagers* passed by. High school students in the '70s could study posters of Saturn and its rings (mine were made from *Voyager* images by the staff at the former Hansen Planetarium in Salt Lake City) and gaze at the *thousands and thousands* of rings that orbit Saturn. Each of the bright segments of the rings seen from Earth can be broken down into smaller and thinner rings—some of them like finely braided strands of hair; some of them, as Rich Terrile would discover, with strange radial "spokes" apparently embedded within them; and some significantly oval-shaped, or eccentric, rather than perfectly circular shapes, seen in more and more detail as the resolution of the images improved. The way the spacecraft's radio signal blinked on and off as it passed through the rings was used to discover that each ring is made up of countless individual blocks of ice, ranging from dust-sized to the size of a house, all orbiting in lockstep with their neighbors. What kept them marching in such orderly fashion?

One answer came from small new moons that were discovered in the images, two of which orbit within a gap in the rings and apparently twist and "shepherd" some of the rings with the push and pull of their combined gravity. If a block of ice in this region starts to wander too far in, one of the shepherd moons will come by and pull it back out; if a house-sized piece of ring in some other region starts wandering away, the other shepherd moon will come by and pull it back in. It was an elegant and completely unanticipated discovery.

Heidi Hammel was an undergraduate at MIT during the *Voyager* Saturn flybys, and she recalls one of her fellow students somehow hacking into NASAs *Voyager* image feed and watching all the images stream by—just like at JPL, but in his dorm room in Boston. When the professors in Heidi's planetary science class got wind of this, they took the class on a "field trip" to this fellow's dorm room to view the images together. "We just sat there, as part of our class, watching the *Voyager* Saturn images come up on this TV screen," she told me. "I remember when the images of the F Ring, the braided F Ring, first appeared. It was all twisted and strange, and we were just staring at it. And I remember that one of my professors, the astrophysicist Irwin Shapiro—I'll never forget this—he was looking at that, and he said, 'Well, that's just not possible. It's not possible for rings to do that.' And we all just started laughing because *there it was* on the TV screen." It seemed like *Voyager* was making the impossible possible everywhere it went.

Another important finding had to do with the composition of the rings. *Voyager* found them to be made of almost pure water ice, with maybe just the slightest hint of a reddish or pink coloration from some unknown but minor contaminant. Earth-based observations in 1977 had discovered thin dark rings around Uranus, and *Voyager* had discovered thick dark rings around Jupiter, which suggested that icy rings get dark over time, maybe from accumulating dust or dark materials from comet and asteroid impacts with the rings. So why would Saturn alone have a gigantic system of superbright, ultrapure rings? The question is part of a bigger debate that is still going on among my planetary science colleagues: how old are Saturn's rings? Their brightness and clean icy nature suggest that they are very young, perhaps having formed from the catastrophic breakup of a large icy satellite just a few tens to hundreds of millions of years ago (that's young to astronomers). However, their highly ordered structure and the amount of mass contained in the rings suggest that any such doomed precursor moon should have been broken up very early in the history of the solar system, when such giant impacts were more common. That suggests that the rings are ancient, and that they keep clean by repeatedly clumping and unclumping with their neighbors as they orbit, continually stirring up "fresh" ice in the process.

Voyager 1 found that not just Saturn's rings but all its large moons are covered by, and mostly made of, water ice. That was not particularly surprising, given the expectations from previous Earth-based telescope observations and the fact that small bodies like moons and asteroids and comets are expected to get icier the farther out in the solar system they form. What was surprising, however, is that the geology of these moons varies so widely. One of the more distant large moons, Iapetus, has a dark hemisphere (the side always leading or facing forward as it orbits around Saturn) that is five times darker than the other, trailing hemisphere. Several of the closer-in large moons, such as Tethys, have global systems of troughs and fractures, suggesting significant but mysterious past active tectonic stresses in their icy crusts. Some of Saturn's moons have preserved a history of intense bombardment over time and are literally saturated with the scars of impact craters formed on their icy surfaces (in the cold temperatures of the outer solar system, ice acts like a rock in many ways). But others have some places on their surfaces that are heavily cratered and other places that are not, as if some process had come along and wiped the slate clean, covering or erasing some of the ancient craters but not others.

Some team members speculated that this process may have been volcanism—but in this case a unique kind of outer solar system process called *cryovolcanism* where the "magma" is liquid water, formed by the melting of the "rock" (solid ice), and erupted through cracks and fissures that let it flow across the surface the same way a rocky lava flow would on Earth. "We certainly did not anticipate the level and breadth of active geologic processes we would find among the outer solar system satellites," Larry Soderblom confessed to me. "We were expecting to explore a bunch of ancient cratered balls with not much else on their surfaces. We should have been smarter." Fortunately, he and the other geologists on the *Voyager* team became smarter very quickly and learned to extend the basic ideas of volcanism—melting of material and transport of the resulting lubricants through cracks or fissures or other underground "plumbing" systems. "Now, because of *Voyager*, we know that no matter where you go in the solar system (or probably in the universe, for that matter) you will always find geological lubricants no matter how cold it gets," Larry went on. "Nitrogen is the melt on Triton, methane on Titan, and ammonia and water likely provide the geological lubricants for many other icy moons. It turns out that volcanism in some form can occur anywhere—whether it is the traditional rocky kind that we know of on the terrestrial planets, or the ultracold cryovolcanic flows and eruptions that we've seen on many of the moons way out there." The specific internal heat sources that could power such volcanoes is still a mystery, although many of my colleagues, like Larry, suspect that tidal heating—like that which drives Io's volcanoes—probably plays a role.

One of Saturn's icy moons proved to be especially enigmatic, as revealed in *Voyager 1* images. Enceladus (pronounced en-CELL-uh-dus) is only about 300 miles across—driving completely around it would be like driving from Boston to Chicago—and yet it is nearly perfectly spherical, has the most reflective surface of all of Saturn's moons, and relatively few impact craters were detected in distant *Voyager* images. This suggested to *Voyager* team members that the surface of Enceladus could be very young, and that, indeed, active resurfacing could still be occurring. In addition, the faint outer E ring of Saturn seems to be densest near the orbital distance of Enceladus, suggesting that this moon might be the source of those ring particles. It was exciting and enigmatic and just plain *weird* for such a small moon to show so much evidence of geologic activity. Enceladus and its neighbor Dione orbit in a 2:1 resonance (like Io and Europa in the Jupiter system), so maybe the same kinds of orbit-related tidal forces heat the inside of Enceladus, melting the icy mantle and causing cryovolcanic eruptions? Hopes were high that *Voyager 2*'s much closer encounter with Enceladus nine months later would provide the evidence needed to solve this puzzle.

Voyager 1's planetary mission ended at Saturn, as the path required for a close Titan flyby, combined with the other constraints, caused the spacecraft's trajectory to be bent upward and well away from any known potential future flyby targets. While *Voyager 2* sped on to strange new worlds, *Voyager 1* settled down for the long journey to the edge of the solar system, and beyond.

If *Voyager 1* hadn't been targeted so closely to Titan and had instead been able to be diverted by Saturn's gravity toward a flyby of Pluto in the 1980s, as NASA had envisioned for some of the original Grand Tour missions proposed in 1969–1970, we would have discovered much sooner that Pluto is a small world with a thin atmosphere, a surface mostly made of nitrogen ice, and orbited by at least *five* moons rather than just the one large one discovered from Earth-based telescopes in 1978. Maybe active plumes or nitrogen-powered volcanoes would have been discovered.

Charley Kohlhase is less sanguine. "I don't think a far-off little, now–Kuiper Belt Object like Pluto, and the trip time and whether you could make it there or not, would ever have beaten out Titan.

Most of the people I knew did not regret giving up on Pluto. We were happy to go with Jupiter-Saturn-Titan for *Voyager 1*, and the smaller Grand Tour Jupiter-Saturn-Uranus-Neptune with *Voyager 2*."

Ed Stone is similarly pragmatic. "Giving up something that you know you can do—Titan—for something where you're not sure you could get there . . . there was no real science controversy about that decision," he recalled. "And we didn't really *know* Titan at all. We probably wouldn't have had a probe landing on Titan if we hadn't focused on it with *Voyager 1*," he added, referring to the successful *Cassini-Huygens* Titan landing mission in 2005. Regardless, we will, hopefully, know for sure what Pluto is really like up close after the July 2015 Pluto flyby of NASA's *New Horizons* spacecraft. Will our biases against the possibilities of life on small, distant worlds continue to be shattered by *New Horizons* as they have been shattered by *Voyager*?

CLOSE CALL

After *Voyager 1*'s successful November 1980 Saturn flyby, all of the team's efforts became focused on planning for *Voyager 2*'s pass through the system in August 1981. With *Voyager 1*'s successful Titan flyby in the can, and no hope of gaining useful additional close-up imaging coverage of Titan from *Voyager 2* because of the thick haze layer, the path was cleared for *Voyager 2* to attempt the Grand Tour. By directing the spacecraft to a very close pass by Saturn, the team could use Saturn's gravity to give the spacecraft a 90-degree turn to aim it toward a 1986 encounter with Uranus, and then, if all went well, possibly on to Neptune in 1989. But getting the

Earth and Moon "Firsts" from Space. TOP LEFT: *Lunar Orbiter I*'s first whole Earth photo from space. TOP RIGHT: *Apollo 8* color *Earthrise* photo from lunar orbit. LEFT: First image of the Earth and Moon together from *Voyager 1.*

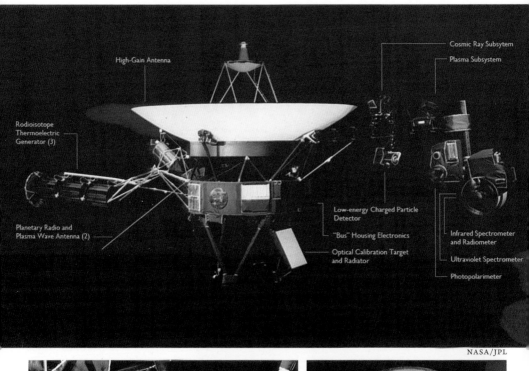

High-Gain Antenna

Cosmic Ray Subsytem

Plasma Subsystem

Radioisotope
Thermoelectric
Generator (3)

Planetary Radio and
Plasma Wave Antenna (2)

Low-energy Charged Particle
Detector

"Bus" Housing Electronics

Optical Calibration Target
and Radiator

Infrared Spectrometer
and Radiometer

Ultraviolet Spectrometer

Photopolarimeter

NASA/JPL

NASA/JPL

NASA/JPL

Voyager and the Golden Record. TOP: Spacecraft and systems/instruments. LOWER LEFT: Close-up of the Golden Record case mounted on the side of the spacecraft bus. LOWER RIGHT: Close-up of the first side of the actual record.

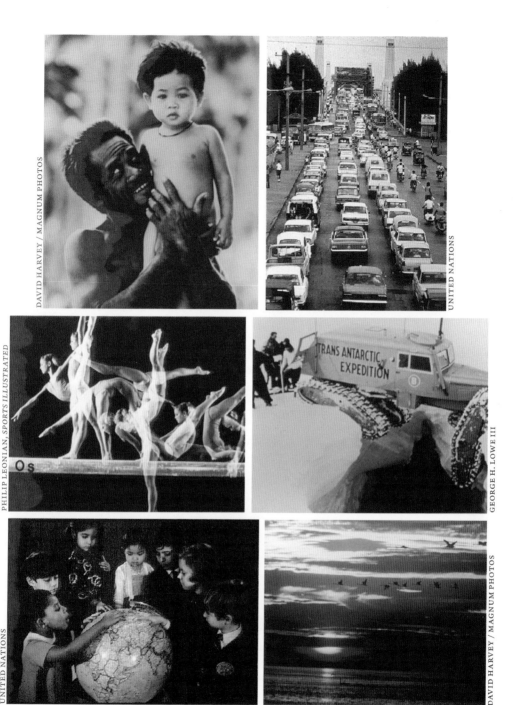

Selections from the *Voyager* Golden Record. TOP LEFT: Golden Record Image 35: Father and child, showing a range of human forms and expressions. TOP RIGHT: Image 102: Rush hour in India, showing many aspects of human transportation. MIDDLE LEFT: Image 71: Stroboscopic photo of gymnast Cathy Rigby, showing the range of human motion over five seconds. MIDDLE RIGHT: Image 108: Stuck Sno-Cat from a 1958 Antarctic expedition, showing that we're not perfect. BOTTOM LEFT: Image 74: Children examining a globe of Earth, with political boundaries. BOTTOM RIGHT: Image 114: Sunset on planet Earth.

DAVID HARVEY / MAGNUM PHOTOS

UNITED NATIONS

PHILIP LEONIAN, SPORTS ILLUSTRATED

GEORGE H. LOWE III

UNITED NATIONS

DAVID HARVEY / MAGNUM PHOTOS

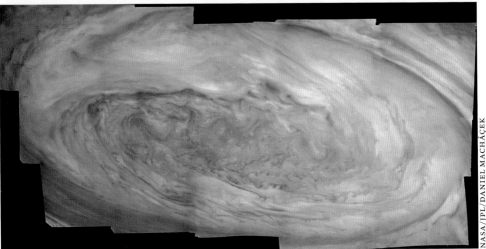

Jupiter: Clouds and the Great Red Spot. Two spectacular examples of modern reprocessed *Voyager* images of Jupiter's Great Red Spot show the regional appearance of this three-Earth-sized storm system (TOP) and a close-up of just the storm itself (BOTTOM).

Europa, Close-up (FACING PAGE). One of the highest-resolution views of Europa obtained during the *Voyager* flybys, this reprocessed version of *Voyager 2*'s closest-approach mosaic shows spectacular examples of the cracks, grooves, and low ridges that imply the existence of a large subsurface ocean underneath this moon's relatively flat, icy crust.

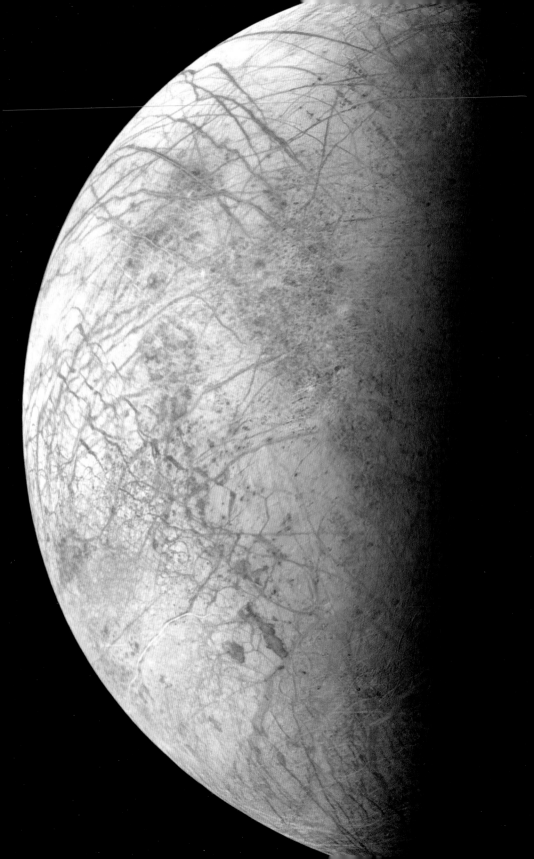

Voyager 2's **Departure from Saturn.** About three days after the closest approach behind Saturn and the major scare from the scan platform anomaly, control was regained of *Voyager 2*'s cameras, resulting in breathtaking, impossible-from-Earth photos like this from beneath the plane of the rings. Modern digital reprocessing of the data helps to bring out additional subtle details in color and structure.

Jumbled Miranda. The highest-resolution images from *Voyager 2*'s flyby of the Uranian moon, Miranda, were digitally reprocessed into this new mosaic of the smallest, and most bizarre, body-orbiting seventh planet. Planetary scientists still don't understand how such abruptly different kinds of terrain could end up all jumbled together like this.

Neptune, Triton, and the Rings. *Voyager* 2 wide-angle and narrow-angle views of Neptune, its faint ring system, its large moon Triton, and simulated views of the background stars at the time of *Voyager* 2's flyby, were combined in this digitally-reprocessed mosaic depicting the last port of call for the mission.

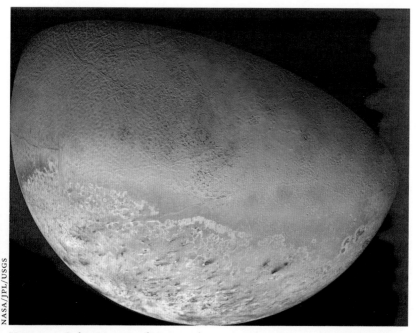

***Voyager* 2 Color Mosaic of Neptune's Large Moon Triton.** The blueish-green "cantaloupe" terrain at center and top of this view consists of ridges and plains of nitrogen ice, while the pinkish, mottled region at the bottom is thought to be a polar cap made mostly of methane ice. Black streaks in the polar cap mark the locations of the geysers detected by *Voyager* scientists.

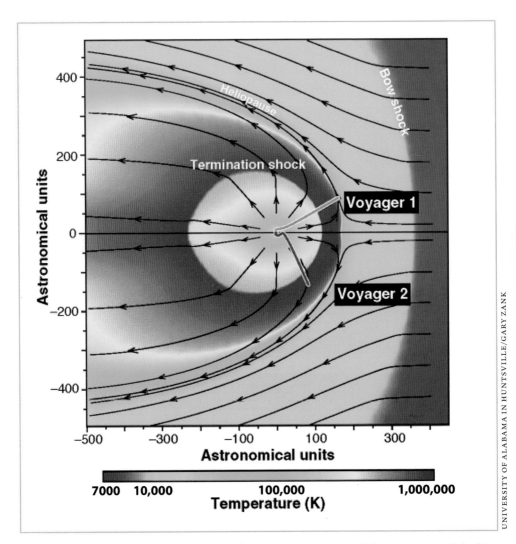

A Simplified Model of the Heliosphere. The sun is at the center of this cartoon model of the "bubble" of solar magnetic field lines (black, radiating out from the sun) and solar wind particles (colored green to red based on their temperatures). The bubble is interacting with interstellar magnetic field lines (black, coming in from right to left) and interstellar particles (colored blue to teal based on their cooler temperatures). The paths of *Voyager 1* and *Voyager 2* as of early 2014 are also shown. The boundary between the green and teal regions is the heliopause, which *Voyager 1* crossed in 2012 and *Voyager 2* is predicted to cross soon.

required course change meant getting very close to Saturn, close to the region where the bright A, B, and C rings orbit around the planet. *Pioneer 11* and *Voyager 1* had shown that the ring particles were (generally) sparsely separated, and that the ring plane could be traversed that close-in to the planet. Still, it had to happen in order to bring about the eventual encounters with Uranus and Neptune, plus it offered close encounters with the enigmatic moons Enceladus and Tethys. The successes of *Pioneer 11* and *Voyager 1* emboldened the team to take the risk—Saturn's rings or bust! So the course was set.

Despite the overall feeling that success at Saturn had already been generally achieved by *Voyager 1*, for some of my colleagues who were involved in the mission, the days surrounding the closest approach of *Voyager 2* to Saturn were among the most harrowing on the mission. Team members who studied the rings, in particular, were nervous. *Voyager 1* had made a rather distant flyby past the rings, but the images still revealed strange waves, ripples, and twisting patterns in the rings that defied explanation ("Rings don't do that!" they said). This was partly because the resolution of the images was too low to show adequate detail. It seemed like some fundamental piece of information about how rings form and evolve, and maybe how waves and wakes form within them, was missing. So they needed to get even closer to the rings with *Voyager 2* to try to understand what was going on. They were nervous, though, because they knew that the closer the spacecraft got to the main rings, the closer and denser the ring particles became, increasing the risk of ring particles impacting the spacecraft.

Even a tiny ring particle, no larger than a speck of silt or a grain of sand, could cripple *Voyager*, which was traveling faster than 31,000 miles per hour relative to the rings. It's the same principle of relative velocity and kinetic energy that explains how you can stop a tractor trailer traveling at 50 miles per hour with a common housefly—as long as the fly is going a million miles per hour! Tiny particles might not seem dangerous, but if they are moving extremely fast, they can carry an enormous amount of energy. Thus, getting closer to the rings offered great scientific potential but also presented great risk—a double-edged sword.

The first half, or inbound, part of *Voyager 2*'s flyby past Saturn was as routine as flying a tight trajectory past a gas giant planet can ever be, and made it possible to capture more of Andy Ingersoll's giant planet atmosphere movies (covering Saturn's northern hemisphere), and distant flybys of Titan and the icy moons Mimas, Dione, and Rhea—all of which had been photographed at higher resolution by *Voyager 1*. It was then that things got . . . interesting.

As the spacecraft fell deeper into Saturn's gravity well and started to speed up (*Voyager 2* was eventually accelerated from 36,000 miles per hour to nearly 54,000 miles per hour by flying a gravity-assist trajectory behind Saturn) and get closer to the planet and the rings, the kinds of maneuvers that the sequencing team had to build into the instrument observations became more and more complex. Specifically, the cameras and other instruments on the spacecraft's scan platform had to be pointed around more rapidly, and over a wider range, than they had been pointed in a very long time. This was a simple result of parallax—the fact that the closer you are to something, the larger it appears to be. Imagine trying to take a picture of the Statue of Liberty, for example, from a mile away on the Staten Island Ferry. The ferry is moving slowly relative to

Lady Liberty, who is far enough away to be leisurely photographed without much need for panning your camera. Now imagine trying to get that same photo from the passenger's seat of an F-15 fighter jet, passing within a hundred yards of the statue and at 1,000 miles per hour! You would have to work fast and avoid craning or whip-lashing your neck to try to catch Lady Liberty in a quick shot without blurring her out. It's the same kind of challenge faced by the *Voyager 2* sequencing team during the closest, fastest part of the Saturn flyby.

All was going well as the spacecraft neared its closest approach to Saturn and—as planned—went into eclipse behind the planet as seen from the Earth. This was the start of a series of carefully or-chestrated observations from a spectacular vantage point inside the Saturnian system that had never before been witnessed. Critical ob-servations would be made of Enceladus and Tethys. Saturn's rings would be viewed edge-on. And the clouds and storms in Saturn's southern hemisphere would be observed up close and personal. The spacecraft would also go into a solar eclipse while behind Saturn, with sunlight shining through its outer atmosphere and scattering through the dense gases as it made its way back to the cameras and instruments on *Voyager 2*. In this way, the spacecraft could measure the chemistry of Saturn's clouds as never before. On its journey, the spacecraft would navigate its way through the plane of the rings, not far from the main rings themselves. *Voyager 2* was on its own for most of this time, hidden behind Saturn, obscuring all possible communications. The intricate series of scan platform slews and camera exposures had therefore been preplanned and uploaded ahead of time. The data would be stored on the spacecraft's 8-track tape recorder and then played back later, when there was time again

151

for dedicated communications back to Earth. A similar scheme had worked flawlessly for both *Voyagers* at Jupiter, and while it would get colder than normal during the eclipse, crossing through the ring plane was a risky maneuver. It didn't help that for about ninety minutes the spacecraft would be out of contact with the Earth. Despite all these dangers, the team was confident that everything would go as planned. Unfortunately, it didn't.

It was after ten p.m. back in Pasadena, and the team knew that *Voyager 2* would be out of communications with the Earth until about midnight. Some team members went home to catch some sleep. Imaging team member Candy Hansen, exhausted from a hectic day of last-minute planning and first-minute data analysis, made it as far as her car in the JPL parking lot and just fell asleep there for the night. Candy says it was pretty common for her and other team members to catch some sleep there at work during the height of the planning and during the encounters themselves. "It was really hard to leave JPL, because you didn't want to miss anything!" she told me. "At each of the encounters I just went out and slept in my car for a few hours. At the Jupiter flybys it was the backseat of my '55 Chevy. By the time of Saturn, I had a little Toyota pickup truck with a camper shell to sleep in. At Neptune I had a van, so sleeping in my car was really comfy by then."

While she and others slept, a skeleton crew, including Rich Terrile, kept watch for *Voyager 2*'s signal to come out from behind Saturn. Rich had invested significant time in planning for the ring-plane-crossing images that would be taken during the spacecraft's pass behind Saturn, as well as the close-up images of Tethys and Enceladus. He couldn't sleep without knowing how those had gone. It is a common theme among the Voyager group: "You didn't want to leave work because something new was just around the corner. The next camera move was going to show something unique. And you wanted to be there to see it, to interpret it. It was an electrifying experience."

When *Voyager 2* did emerge from behind Saturn, there were some cheers and perhaps even a few unconfirmed Champagne corks popped among the small group still on shift. But soon it was clear that something had gone wrong. "All of a sudden, the pictures stopped coming. They were sort of frozen," recalls Rich Terrile. "Oh my gosh, we've got a big problem here." Telemetry showed that a series of hardware and software errors had occurred while the spacecraft was behind the planet, and that the spacecraft was not what the engineers call "healthy." Something had happened to cause the preplanned sequence to stop taking data.

"It was an amazingly frustrating, shocking, kind of 'What do I do? I'm totally helpless' kind of experience," Rich Terrile recalls, "where suddenly the spacecraft is *not* doing something when it's *supposed* to be doing something. You don't have any idea what's going on, and it's a billion miles away." He remembers feeling as if he went through the various stages of grief (shock, denial, anger, and so on) compressed into a few minutes. "This can't be happening!" he felt.

Ed Stone had been catching up on a little sleep at home that night as well, but got up early to head back to JPL to be in a morning East Coast media interview about *Voyager*'s results and status. "I think it was probably four in the morning," he recalls, "and I showed up and they said"—he whispers—"'Scan platform—it's stopped!' So I had to go on this live TV interview and talk about it, even though I'd just heard, and didn't know anything!" No one else knew much yet either. Ed remembers that the team's primary feelings were of real concern for the spacecraft and of course the future of the mission. "We'd had a wonderful Saturn encounter, so Saturn wasn't really the issue. The issue was, is this a spacecraft that's going to get to Uranus and Neptune?"

At first, many people's thoughts turned to the ring-plane crossing. Maybe *Voyager 2* had rammed into a chunk of ice or dust floating among the outer rings? The Plasma Wave Subsystem instrument's PI, the late Fred Scarf of TRW, reported to the *Voyager* science team the day after the scan platform anomaly that his instrument had detected activity "a million times the normal energy level very close to the time of the ring-plane crossing." Scarf played a cassette tape recording of the "sounds of the ring plane crossing" and hypothesized that they were detecting bursts of energy from the high-speed impacts of small dust grains with the spacecraft. As reported by *Voyager* imaging team member David Morrison in his almost minute-by-minute account of the Saturn flybys, "The quantity of such impacts was truly staggering—thousands upon thousands, not just at the moment of ring-plane crossing, but extending for several minutes on either side. The roaring sound of these impacts on the tape that he played, sounding almost like a hailstorm striking a tin roof, sent chills down the spines of the seventy-five scientists attending the meeting. But did this unexpected plasma activity really have anything to do with the scan platform failure? No one could tell."

Had it been too much of a risk to take? If *Voyager* had crashed into a larger-than-normal (a micrometer or two in size) ring particle, it must nevertheless have been an extremely small particle, because the spacecraft's trajectory when it came out from behind Saturn was still exactly on target for where it was supposed to be. Still, even tiny impacts with ring particles could cause localized problems with some of the spacecraft's instruments and subsystems, so the possibility was taken seriously as a potential culprit.

Another possibility was that the scan platform mechanism had failed because of the larger temperature contrast encountered when *Voyager 2* went behind Saturn. As JPL's testing of the flight scan platform showed, as its mechanisms and components were used, they heated up. Meanwhile, the outside temperature of the spacecraft itself got colder than it had ever been while it was passing though the eclipsed darkness of Saturn's shadow. It was possible that wear and tear and "normal" aging of the scan platform's gears and drive shafts and lubrication in the deep-space environment could respond to sudden, more extreme temperature changes in unpredictable ways—so that hypothesis had to be taken seriously too.

Ultimately, most team members believe that the reason for the failure was the sheer complexity of what the spacecraft and its mechanisms, especially the scan platform, were being asked to do— pointing the cameras and other instruments quickly and over wide angles from target to target during the high-speed, high-parallax closest approach. "We were cranking that thing around a lot," recalls Rich Terrile. Overwork of the gears within the scan platform, in particular, was treated as a serious possibility, and it was a hypothesis that could be directly tested using the flight spare scan platform that was back on Earth, at JPL. Based on that work, the spacecraft engineers quickly devised some tests that could be done on *Voyager 2*'s actual scan platform, to try to recover at least some science as the spacecraft quickly receded from Saturn. Commands were sent to *Voyager* to move the scan platform back toward Saturn, but in smaller, slower steps than it had been previously commanded. Easy does it. . . .

"If it responded properly, the first picture would be received at 5:38 p.m.," reported Dave Morrison in his diary of events for August 28, 1981. "All over JPL, people gathered around the television monitors as 5:30 approached. The three-minute delay between pictures always seemed long, but in these final moments time seemed almost to stand still. Then, at last, the critical picture began to be displayed, line by line, on the screens. And there it was! Clearly visible was a bit of the planet with the rings, now seen for the first time on their dark side, arching up across the field of view. A vast collective sigh of relief was expressed all around the Lab." A work-around had been found. Some data were lost but the spacecraft, camera, and scan platform were responding well to triage. We might get to Uranus and Neptune after all!

What exactly happened when *Voyager 2* was behind Saturn remains a mystery. The favored hypothesis for the scan-platform failure behind Saturn, based on generally good simulations and reproduction of the failure in the spare scan platform at JPL, is overwork and overheating from high-rate slew, and the seizing of a small gear on a tiny shaft within the mechanism. Still, it is hard to let go of the coincidences associated with the spacecraft's cold pass through Saturn's shadow, and especially of the daring plunge through the ring plane so close to the planet itself. In the years that followed, mission planners would wonder if future missions, such as the planned joint NASA/ESA *Cassini* Saturn orbiter and *Huygens* Titan lander, should take a more cautious approach to studying the Saturn system.

In 2004 the author Richard Hoagland, who covered the *Voyager 2* Saturn flyby for the media and JPL and who by that time had become somewhat famous (or infamous, depending on whom you talked to) for identifying the so-called Face on Mars from NASA *Viking* orbiter images, accused NASA of sending the impending *Cassini* orbiter on a "suicide mission" to Saturn, knowing the fatal risks that *Voyager 2* had barely avoided when crossing through the ring plane. Or, Hoagland asked in a more ominous tone, could NASA be covering up some remarkable, revolutionary physics discovery that *Voyager 2* secretly made during that pass behind Saturn in 1981? About NASA's *Cassini* plans he asked, "Is the spacecraft actually more prepared for such a 'secret mission'—including being able to survive its highly dangerous ring plane crossings—than we have been publicly led to believe . . . ?" Well, secret mission or not, *Cassini* survived its own daring plunge through the ring plane of Saturn in July of 2004, and, building on the discoveries by the *Voyagers*, has gone on to orbit the ringed gas giant more than two hundred times, revolutionizing again our view of the planet, its moons, and its rings, as the *Voyagers* had done twenty-five years earlier.

Voyager 2's troubles on the far side of Saturn were an acute reminder to science team members of the risks inherent in space exploration, and of the fact that there are often psychologically painful costs associated with those risks. Dave Morrison reported that "a sense of gloom pervaded the Imaging Science area. Everyone realized that these would be the last close-up views of Saturn or its satellites they would ever receive—what Rich Terrile called 'our last best data.' It was like watching each labored breath, waiting for the sick scan platform to expire." Candy Hansen remembers the sad scene. "That particular day after the anomaly, my boss and fellow imaging team planner Andy Collins and I commandeered the browse room and sadly went through every image that had been taken but missed its target. . . ." Linda Spilker was similarly distraught. "I remember feeling a tremendous sense of loss of all the science observations that I had helped to plan," she said.

While many high-resolution images of Saturn and its rings were successfully acquired before the scan platform stopped moving, particularly painful for the planetary geologists on the team was the fact that the platform had failed just prior to the highest-resolution imaging of Tethys and Enceladus. "Instead of satellite images," reported Morrison, "only a blank screen appeared." Given that these would be the final planned planet or moon images of *Voyager*'s nominal Jupiter-Saturn-Titan mission, the ability of the spacecraft to potentially continue on to successful encounters with Uranus and Neptune was now in grave doubt. And given the fact that budget cuts were decimating the planetary exploration program in general (with no new outer solar system missions then in the works), an anonymous team member also watching the parade of blank images march by the screens lamented out loud that these could be "the final images of the planetary program."

In 2005, *Cassini* Saturn orbiter photos of that tiny moon finally explained the enigmas noted by the *Voyager 1* team back in 1980: Enceladus is indeed geologically active, and plumes of water vapor mixed with organic molecules were seen spewing out of a series of "tiger stripe" cracks in that moon's south polar crust. Heated by tidal forces, the interior of tiny Enceladus is partially molten ice— liquid water—that makes its way out via fractures in the crust and helps supply the particles that make up Saturn's faint E ring. Instantly, Enceladus was vaulted to a prime position, along with Mars and Europa, on the solar system's short list of astrobiology "hot spots"—places where liquid water, heat sources, and organic molecules could have made the environment habitable for life as we know it. Maybe *Voyager 2*'s scan platform was dinged by one of these pieces of Enceladus back in 1981? Maybe the Enceladusians just weren't ready to be discovered yet?

In any case, *Voyager 2* had survived mostly unscathed and still on course, so JPL *Voyager* engineers and science team members did what they often do best: they improvised, and devised work-arounds, to solve a problem by remote control. "Fortunately, we had five years to sort it out," said Ed Stone. Through testing of the spare JPL scan platform, strategies were devised to enable its effective use in the future (though never again at high speed), if *Voyager 2* could survive to Uranus and beyond. In fact, the engineers devised a way to use the *entire spacecraft* as one big scan platform, in what Torrence Johnson called an "anti-smear campaign," by using gentle puffs (each with a thrust of about 3 ounces) from its attitude-control jets to impart a slow, gentle roll. "We always knew that the light levels at Uranus and Neptune would be low," Charley Kohlhase says, "so you'd have to use longer exposures. If you wanted to get good pictures, you'd have to avoid smearing them. The flight team knew that by firing a few pulses from the attitude-control thrusters, you could put the spacecraft in a gentle roll in a direction that compensated for the apparent motion of a target." While it was known early on that *Voyager* could be controlled this way, "We just didn't deal with it," Charley told me, "until we had to." Luckily, in some ways, the years it would take *Voyager 2* to travel out to Uranus—twice as far away from the sun as Saturn—would give the team plenty of time to test these strategies and to prepare for an encounter with the truly unknown.

6

Bull's-Eye at a Tilted World

WHEN I WAS a kid, we'd take long summer family car trips through New England or up and down the East Coast. We had a variety of interesting cars (a different one seemingly every week—my father ran a junkyard and auto-repair shop), but the ones I remember the most were the kind of "family truckster" station wagons made famous in movies like Chevy Chase's *National Lampoon's Vacation.* My sisters and I would romp and play in the backseat, in the seat wells, or in the "back back" among the luggage and spare tire, inevitably playing games that somehow involved us hitting each other. Good times. I'm not sure that any of these cars even *had* seat belts—not that we'd have known how to use them. It was a different age, to be sure.

Members of the *Voyager* team found their own ways to pass the time during the long interplanetary cruises between flybys. Many team members planned vacations to coincide with the end of each planetary flyby as a way to blow off some steam from the stress (and sometimes, frustration) of planning for and acquiring "one-time-only" observations as the *Voyagers* sped past. Such respites were usually short-lived, however, because once back from vacation, the stresses of another impending flyby would once again slowly start to accumulate. Some team members took on new jobs with new projects, either using their *Voyager* experience to land a better position at JPL or elsewhere, or were forced to seek alternate work at JPL because the project's budget was being trimmed yet again.

Between Saturn and Uranus, Candy Hansen took a two-and-a-half-year leave of absence from JPL and went to work as a science and operations liaison at the German Space Operations Center in Oberpfaffenhofen, learning how other countries operate their spacecraft and exploring Europe in a little camper during the generous German summer vacation seasons. Some team members also retired during the course of the mission, having started on *Voyager* when it was just a gleam in their eyes fifteen or twenty years earlier. Some team members even decided to time their weddings and the births of their kids to coincide with "downtime" on *Voyager*, which of course could be predicted years ahead of time. Between Uranus and Neptune, Candy Hansen had a daughter and went back to graduate school at UCLA to get her master's degree (followed by her PhD after the Neptune flyby). Linda Spilker recalled how she had to interweave her personal life with her professional life, including starting her family. "I tell my daughters, Jennifer and Jessica, that their births are based on the alignment of the planets, and I mean that! There was about a five-year window between the *Voyager*

Saturn flyby in 1981 and the Uranus flyby in 1986. It was in this window that I had both of my daughters. Other *Voyager* moms made similar choices and our kids grew up together." This is sort of a modern-day version of astrology—people's lives being dictated by the positions of the planets. Those *Voyager* kids, in particular, were born when they were born only because the outer planets happened to align every 175 years or so, the most recent alignment happened to occur when our species (finally) had the technology needed to exploit it, and at least one of their parents happened to have a job that also depended on the alignments of those planets. Voilà! Astrology works.

The years between the *Voyager* Saturn and Uranus encounters were formative years for me as well, as I graduated high school and moved out west to become an undergraduate at Caltech. I had to take the required math and physics classes, but for my electives I dabbled in astronomy, engineering, and planetary science. Many faculty members in the Division of Geological and Planetary Sciences were directly involved with the *Voyager* missions, and in class they would sometimes take us students on joy rides to test-drive some of their scientific models or explanations for the images and other data that had only relatively recently been taken by the *Voyagers* at Jupiter and Saturn. Viewgraphs, sometimes hand drawn, 35mm slides, even film-loop movies would be bandied about while the professors argued with one another and with the grad students in the class, and the grad students argued among themselves. Most of us undergrads in the classes were blown away by the jargon and the complexity level of the material, but still, we were in hog heaven. Here was the process of science and discovery, happening right in front of us! We struggled to get B and C grades but felt good about passing at all.

I remember the debates about Uranus were particularly poignant, because the professors and grad students were so keenly attuned to the fact that this basically unknown planet and its rings and moons were going to be almost magically revealed to us all by *Voyager* 2 very soon. There was a lot of speculation about what *Voyager* would see, and about how this smaller, more blue-green giant planet would compare to its larger cousins, Jupiter and Saturn. Are all giant planets the same? they wondered. Are there patterns in the ways that giant planets behave depending on their size? Or on their distance from the sun? Out at Uranus, sunlight is about four times fainter than at Saturn, and more than thirteen times fainter than at Jupiter. What difference would that make?

BIG SCIENCE

Uranus (pronounced by professional astronomers as YUR-uh-nus rather than the way eleven-year-old boys pronounce it, while giggling, your-ANUS) is special among the planets. To start with, it is the first planet that was discovered by telescope. The ancient astronomers of Greece, Persia, China, and Babylon didn't know it existed. It is so far away from us, orbiting at an average distance of almost eighteen times farther than the distance between the Earth and the sun, that it is usually too faint to be visible to the naked eye. Even if it were naked-eye visible, it would still have appeared starlike and would have been moving so slowly relative to the real background stars (taking eighty-four Earth years just to orbit the sun once) that it would not have been recognized as a particularly different or special star. There are reports of some very astute ancient astronomers, and perhaps even Galileo Galilei, the inventor of the first astronomical telescope in 1610, having seen a starlike object where Uranus should have been at the time. But no one recognized it as a *planetes asteres*, a wandering star.

At least, not until March of 1781. That's when the German-born English musician and astronomer William Herschel (1738–1822) single-handedly doubled the size of the solar system. Herschel is one of those almost mythic characters from the early history of Western astronomy. Like Isaac Newton in the century before him, Herschel was a prominent polymath—a person with many different kinds of technical and academic skills. He focused much of his energy on, and made his living by, composing and performing music (twenty-four symphonies, fourteen concertos, and many other pieces) and also dabbled in astronomy, optics, and other areas of science. This dabbling eventually led him to design large mirrors and telescopes (based on Newton's design) and to begin making systematic observations of the night sky. Soon enough, using a 6.2-inch-diameter, 7-foot-long telescope from his home observatory in Bath, he noticed an object with a "nonstellar disk" moving ever so slowly through the stars. Instead of appearing as a point of light like the stars, this object had a circular nature to it. Collaborating with others, Herschel was able to show that the object moved on a planetary path, well beyond the orbit of Saturn.

Herschel's first instinct was to name the new planet Georgium Sidus (George's Star), after his king and patron, George III. In that regard he followed in the well-established footsteps of previous astronomical discoverers who wanted to (for lack of a more fitting term) kiss ass by naming their findings after the people who paid their bills. For example, in 1610 Galileo wanted to name the four bright dots that he discovered orbiting Jupiter the Medician stars, after his patron and funding source Cosimo II de' Medici (Grand Duke of Tuscany) and Cosimo's three brothers. Thankfully, astronomers decided on Io, Europa, Ganymede, and Callisto instead. Anyway, no one (except perhaps King George) liked Herschel's proposed new name, especially the rival French, who referred to the new world simply as Herschel. Astronomers eventually settled on Uranus, the Greek god of the sky. A few years later, using an even larger (and more cumbersome) telescope of 18.5 inches in diameter, he discovered two moons orbiting Uranus (later named Oberon and Titania by Herschel's son, John, who also became a prominent astronomer), and shortly thereafter he discovered two more moons around Saturn—Mimas and Enceladus. All this while he continued to compose and perform music and while, "on the side," he discovered infrared radiation, contributed to the study of Mars and the other planets, and made biologic observations with his microscopes.

Herschel was motivated and accomplished for sure, but like many others with such broad interests and talents, his didn't do it all alone. Indeed, one of his most capable assistants was his sister, Caroline Herschel (1750–1848). She aided with his music and toiled at the telescope, contributing to new telescope designs, polishing the mirrors, and performing observations of her own. At a time when women had few opportunities in science and academia, Caroline became an accomplished astronomer in her own right, with an outstanding record of instrument building, comet discoveries, and important work cataloguing faint stars. While she never became a full member of the all-boys club that was the Royal Astronomical Society back then, she came as close as any woman had (or would for almost a century more), and though she deserved more, she was elected to "honorary" membership in recognition of her contributions.

The way that the Herschels—father, sister, son—had to work together and with other science and engineering/technology colleagues of the day to make their groundbreaking discoveries seems like a great example to me of one of the earliest cases of the slowly growing phenomenon of "Big Science." Science, perhaps especially astronomical science, generally begins with highly motivated individuals making careful observations, or working out interactive theories, essentially on their own or with just a few select others. Notable examples from the history of Western astronomy include pioneers like the sixteenth-century Polish astronomer Nicolaus Copernicus; Danish observer Tycho Brahe and German astronomer Johannes Kepler working together in the late sixteenth century; the late-seventeenth-century English physicist Isaac Newton; and of course the first loner at the telescope, Galileo Galilei, in the early 1600s. But the history of individual, or "small," science pushed forward mostly by key individuals goes much farther back in time and crosses many cultures, including notable Greek, Arab, Persian, Chinese, Indian, and other thinkers. The idea of collaborative science was generally more rare, although there were some important and profound early advances in math, physics, and astronomy made by larger groups working together, such as the astronomer and mathematician Nasir al-Din al-Tusi and his thirteenth-century research team studying planetary motion at the Maragheh Observatory in Iran. Or the academics working together on early forms of calculus to propose some of the first models for a sun-centered universe in the sixteenth-century Kerala school of mathematics in India (influencing, in their writings, another loner astronomer in Poland, the previously mentioned Nicolaus Copernicus). Or Harvard astronomer Edward Pickering's early-twentieth-century group of mostly female "computers" toiling through enormous telescopic data sets to work out the modern basis for the classification of stars.

As technology has advanced, and the breadth of knowledge required to understand, utilize, and improve that technology has expanded, it has become more difficult for individuals, or even small groups of people, to define the cutting edge of science, and especially space science. Projects like the *Apollo* moon landings, the *Voyager* missions, the Hubble Space Telescope, or rovers on Mars require detailed theoretical calculations (to determine orbits, or to estimate camera exposure times in preplanned sequences, for example), advanced engineering technologies (such as new materials, new kinds of instruments, new software for communications and commanding), and clear scientific goals based on the most recent laboratory, telescopic, and computational discoveries. It is simply not possible for small groups of people to pull off projects of this scale, and so big teams with wide ranges of expertise are needed to design, build, and operate experiments, and to process and interpret the results. Big Science.

ROLLING WITH THE PUNCHES

When William and Caroline Herschel discovered Oberon and Titania, they and others couldn't help but notice that those moons (and others found later) are spinning around Uranus in a vertical plane, like the wheels on a car rather than like a record on a turntable. The spin axis of Uranus is tilted on its side, at an angle close to

90 degrees relative to the rest of the planets. Uranus rolls around the sun, rather than spins. What's up with that? One of the main goals of the *Voyager 2* flyby of Uranus was to try to find some clues to answer that question.

To try to come up with some ideas ahead of time, astronomers tried to understand why all the other planets spin like tops with their poles pointed roughly (within 20 to 30 degrees or so) perpendicular to the equator of the sun. The prevailing idea is that the sun and all the planets formed some 4.65 billion years ago from a condensing, spinning cloud of gas and dust. The cloud must have been spinning counterclockwise as viewed from above the north pole of the sun, because that's the direction that the sun spins on its axis and that all the planets orbit around the sun. The notion that the planets formed from a *disk* of gas and dust helps to explain why their poles are pointed north-south: their equators are all forming within the plane of that relatively flat disk. But Uranus is an oddball: its equator is tipped over by 90 degrees. For part of its eighty-four-Earth-year trip around the sun, the north pole of Uranus is pointed right at the sun, and the entire southern hemisphere is dark; forty-two Earth years later the situation is reversed in southern summer, with the northern hemisphere in the complete darkness of polar night. In between, near the spring and fall equinoxes, sunlight falls on both hemispheres. A planet's tilt determines the intensity of its seasons: Earth's 23.5-degree tilt produces extreme seasons with constant sunlight or constant darkness for people or animals that live above the Arctic (or Antarctic) circles; Jupiter's tilt is near 0 degrees, and so despite being superlative in many things, it has no seasons. Uranus has the most extreme tilt and thus the most extreme seasons, with its own Arctic circle falling very close to its equator.

How did Uranus get this way? No one knows for sure, but one popular hypothesis is that Uranus formed "normally" like the rest of the planets, but early in the history of the solar system it was knocked on its end, tipped over, by a giant grazing impact with another large terrestrial planet or a small gas giant. That impact would have essentially melted both bodies, but if a newly forming condensing cloud of post-impact gas and dust had started to spin vertically because of the force of the impact, that orientation could end up being the new tilt for a newly formed (potentially merged) planet.

It sounds outrageous and ad hoc . . . because it is. In general, scientists don't like to invoke special one-time events like this to explain the world(s) around us, but sometimes, to paraphrase the famous Sherlock Holmes, if you've ruled out the impossible, and all you're left with is the improbable, that's probably the right answer. Only recently in planetary science has the idea of giant impacts as major agents of planetary change become more widely accepted. Indeed, despite the crazy sound of it, the idea of a giant grazing impact between the very young Earth and a Mars-sized protoplanet is the best explanation of the formation of our moon, based on *Apollo* samples and analysis of the Earth's interior composition.

So maybe giant impacts aren't that crazy after all. Maybe there would be something about the planet's magnetic field lines (if the planet had a magnetic field—*Voyager 2* would find out!) or interior structure that would prove to be the smoking gun that supported *some* model for why the planet is tilted over. Maybe, maybe not.

"We thought a lot about that," Ed Stone says. "But we had a terra-centric view that was limiting our considerations: We *assumed* that like all other magnetic fields that we'd seen at that time that the magnetic pole would be near the rotational pole. So we were expecting to see a unique situation where the solar wind was directly impinging on the planet's south magnetic pole. On the Earth, that's the 'funnel' where particles come in, and it's a really interesting place." The "funnel" Ed is talking about is the convergence of the Earth's magnetic field lines near the north and south poles. The field lines coming together act to concentrate the high-energy solar wind particles that are streaming along those lines, increasing their density and the intensity of their interactions with the Earth's atmosphere. This is part of the reason that Alaskans and Canadians and Scandinavians (and Antarctic penguins) see such intense and beautiful auroral displays—the funnel concentrates the energy, and the aurora is one of the ways that that energy is dissipated. I can imagine Ed and others wondering if they would witness similarly spectacular auroral displays at Uranus. The reality, however, turned out to be quite different.

The crazy tipped-over geometry of Uranus meant that instead of a relatively leisurely, multiday tour past the planet's moons and rings like at Jupiter and Saturn, *Voyager 2* would instead be flying a banzai-like bull's-eye trajectory, piercing through the Uranus system like an arrow flying through a target at over 51,000 miles per hour, with only about ten hours to conduct all the needed close-up observations. The *Voyager* navigation team needed to aim for a specific closest-approach point within just a few diameters of the planet in order to make the spacecraft pass through the shadow of Uranus (to measure the atmospheric composition and structure) and to give *Voyager* the needed gravity-assist tweak to send it on to

Neptune and complete the hoped-for Grand Tour. Passing that close to the planet and its bull's-eye pattern of moons meant that mission planners could only try to tweak the timing of the flyby so that *Voyager* would pass close to the innermost moon, tiny Miranda. It was unfortunate that the other moons couldn't be studied so closely—just bad luck because of the geometry of the flyby. On the other hand, unbeknownst to the team, Miranda would turn out to be the most interesting of the five large icy satellites of Uranus.

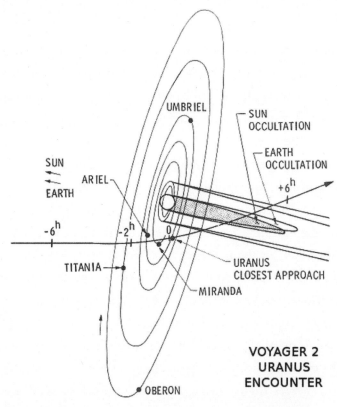

Bull's-Eye. *Voyager 2* flyby trajectory past Uranus. *(NASA/JPL)*

There was another, more serious, problem that the *Voyager* team had to solve in order to make the Uranus flyby a success: image smear. Four times less sunlight at Uranus than at Saturn meant that the cameras and other instruments would have to expose their pictures four times as long to get the same image quality. But this was a fast flyby—less than half a day, and the spacecraft had picked up an extra 18,000 miles per hour of speed by the gravity assist at Saturn. The team had already noticed some small amount of image smear in the Saturn images compared to the Jupiter images (where the sun was brightest and exposures shortest among the entire Grand Tour). Some of the smear seemed to be caused by the jolting starts and stops of the tape recorder, which shook the spacecraft a little bit during long-exposure photos. Leaving the shutter open even longer would mean that the images would be smeared out even more. Instead of seeing crisp views of new worlds, blurry, streaky photos would be taken. The team had to find a solution, and indeed they did: *Voyager 2* was almost completely reconfigured and reprogrammed between Saturn and Uranus, becoming a faster, smoother photographer and a more high-tech spacecraft.

First, they taught the spacecraft to move from target to target—to slew and change its "attitude" or orientation—much more slowly and smoothly than it had done previously, using tiny puffs of well-timed thrusts from the hydrazine attitude-control system. Then they taught the spacecraft to anticipate, and to compensate for, the tape-recorder jolts using the same attitude-control system. The most elegant new mechanical trick that they taught *Voyager*, however, was something called *image motion compensation*. Even though the team had several years to diagnose and correct for the problems that *Voyager 2* had with moving its science instrument scan platform too quickly during the most rapid-fire part of the Saturn flyby, mission planners didn't want to take the chance of the same thing happening again during the similarly rapid Uranus bull's-eye flyby sequences. So they restricted the planned motion of the scan platform and instead taught the *entire spacecraft* to gently pirouette (or at least partially pirouette) in the direction *opposite* its motion relative to Uranus and its moons when it was imaging them. From the viewpoint of the camera onboard the spacecraft, this would make those bodies appear to zip by more slowly. It was a ballet.

Voyager's brains—its main computer and its backup computer—also got major overhauls to prepare for the Uranus flyby. The main computer was reprogrammed to send its images immediately to the backup computer to process and compress them there instead of in the main memory, speeding up the rate that *Voyager* could take pictures by 70 percent. The computer was also taught how to use an experimental new data-encoding box that was on the spacecraft but that hadn't been used at Jupiter or Saturn. "Data encoding" is the process of converting the images and other data into the compressed string of ones and zeros that would be broadcast by radio back to Earth. The new encoding scheme would be more efficient, more robust, to the weaker signal levels from Uranus in that it would enable JPL communications engineers to better reconstruct the original data even if some "packets" of the radio signal were lost or corrupted. It was also predicted to work better than the default encoding routine for the low light levels and dark moons and rings expected at Uranus.

Even the JPL Deep Space Network team got in on the game, increasing the sensitivity of their receivers to deal with *Voyager 2*'s weaker and weaker radio signals as the spacecraft sped farther away, and adding the capability to communicate with *Voyager 2* from the

Parkes Radio Telescope in Australia (the one made famous by successfully broadcasting Neil Armstrong and Buzz Aldrin's *Apollo 11* lunar landing live to the world in 1969) during the most critical part of the Uranus flyby—which would be best visible from the Australian DSN stations. As Ed Stone wrote in one of the early scientific papers describing the *Voyager 2* results at Uranus, "That all of this worked so well testifies to the high level of expertise and the spirit of teamwork within the *Voyager* project and supporting organizations." Big Science.

GEOGRAPHY REINVENTED

Voyager began observations of Uranus and its surroundings in November of 1985, while still more than 6 million miles from the planet. Every so often I would swing by Ed Danielson's office at Caltech, or by one of the conference rooms in Millikan Library on campus, to catch a glimpse of the planet on one of the monitors there that was echoing the steady stream of images being sent to JPL. Students and staff would often gather around the small black-and-white displays, leaning in and squinting to read the telemetry text information also displayed in tiny type along with each picture. We could monitor the increasing apparent size of the planet by comparing it to the distance between the black dots called *reseau marks* that were etched onto the camera's lens and which thus appeared in every image. In some images, we could even start to see the faint doughnutlike ghost from out-of-focus dust specks on the lens. Heading toward the end-of-year holidays, this new greenish-blue world was starting to get big in the headlights.

I went back to Rhode Island to visit my family over that holiday season and felt acutely, almost completely, detached from *Voyager*. Computer science and technology might have been slowly burgeoning at NASA, but the Internet was still just a small academic network among universities and select government labs—a far cry from the publicly available web of infinite knowledge and connectivity that we take for granted today. TV and newspaper media coverage of the flyby was rare, right up until the few days around the closest approach in late January. Once I got back to Pasadena, I used my magic *Voyager 2* team badge that Ed Danielson had gotten for me to get myself over to JPL as often I could, sometimes driving my 1963 Ford Galaxie 500 convertible (white, "three on the tree") the seven miles between campus and JPL, sneaking into the visitors' daytime parking lot, and a few times skipping out of class to bum a ride with Ed when he was making the trip. As it got closer to encounter day, parking became scarcer, with big TV vans and buses taking up space in the visitors' lot and outside Von Kármán Auditorium, where the press conferences were being held. Even when I had to park more than a mile away, it didn't matter. I couldn't keep myself from running all the way to the lab.

During the days right around closest approach on January 24, security was ramped up considerably at JPL. Guards were posted in the science work areas of Building 264. Limits were placed on the number of people allowed at once in the operations rooms, for fire code reasons, I'm sure, and also because of the need to preserve a relatively peaceful environment for performing tactical calculations and making weighty decisions about the flyby. For example, in late December, a faint new moon was found orbiting Uranus in some of the *Voyager 2* approach images. It was imaginatively dubbed 1985U1 (and later officially named Puck after the airy sprite in Shakespeare's *A Midsummer Night's Dream*). Mission planners realized that one of the preplanned images of Miranda could be reprogrammed to get a decent, though distant, view of this small new world. There was work to be done.

I wasn't really doing any of the important work. Rather, I was a gofer (go for this, go for that) for Ed or other team members who needed copies made, a print retrieved from the image-processing lab, a cup of coffee, or a pizza delivery. I tried to avoid being tossed out for bumping into something, or even for breathing too loudly, or too much. I'm sure I had a dopey grin on my face most of the time.

We all knew that Uranus didn't have strong, high-contrast cloud bands like Jupiter, or bright rings like Saturn, but I think many of us ended up being quite surprised at just how *bland* the atmosphere turned out to be, even at high resolution, when *Voyager 2* sped past. The blue-green color was strikingly different, to be sure—caused by methane in the upper atmosphere that absorbs mostly red light from the sun and reflects the rest. Weak, broad bands of slightly brighter or darker tones could be seen at some latitudes, and some occasional small white clouds—almost like smeared-out thunderheads—would pop into view and zip past now and then. But this giant planet was definitely a different flavor, a different beast, from the ones we had seen before.

Ed Stone and his "fields and particles" colleagues were delighted (though, he confessed, "not surprised") to discover that Uranus does indeed have a strong magnetic field, and were even more delighted when that magnetic field turned out to be pretty weird. The fields that *Voyager* had found around Jupiter and Saturn were much stronger, and sort of what the science team had expected. Deep within the interiors of those giant planets are mega-Earth-sized cores of hydrogen compressed to such high pressures and temperatures that the gas acts more like a metal, easily conducting electricity. We know from our own planet's interior that spinning, electrically conducting cores of planets (such as Earth's partially molten iron core) can generate a magnetic field. And the shape of that field is sort of like the shape that iron filings take when they are exposed to a regular bar magnet—the field lines orient along north-south magnetic "poles" that come close to lining up with the north-south spin axis of the planet. That's why a compass can tell you which way is north.

At Jupiter and Saturn, the magnetic fields looked basically as expected—like the fields that would come from giant bar magnets placed at their centers. The field lines get warped and bent and "swept back" by the solar wind, streaming out almost cometlike into a long magnetic "tail" (called a *magnetotail*) that points away from the sun. In fact, Jupiter's magnetic field and magnetotail are so enormous that if we could see it with our naked eye it would be five times larger than the full moon in our night sky—making it the largest single structure in the solar system except for the sun's own magnetic field. Saturn's field is not as large but is similarly impressive. By comparison, *Voyager* measurements showed that the magnetic field of Uranus is quite different, probably because of the crazy tilted way the planet spins. It's almost like there's a giant bar magnet down below the clouds—meaning that there is a rapidly spinning, conducting material of some kind—but Uranus's is not located in the center of the planet. Instead, it appears to be offset to one side about one-third of the way to the cloud tops. And unlike Earth's, the magnetic-field axis is not even close to being lined up with the planet's edge-on spin axis; instead, it's tilted at an angle of about 60 degrees from the orientation of the planet's spin. Your compass won't do you much good there.

"The combination of the planet's spin axis tilt, and the offset, tilted nature of the magnetic field, combined to make the angle of the field relative to the solar wind really not much different than all the other magnetic fields that we've encountered," says Ed Stone. "But what was different was that, unlike the others, the tail end of the Uranian magnetic field is a spiral, because it's being wound up by the planet's tilted-over spin." The solar wind warps and sweeps this bizarre and unexpected structure downwind in a unique way.

One of the possible implications of the planet's strange magnetic field is that the inner core of Uranus, which *Voyager* gravity data showed is rocky and icy and about the size of the Earth, is not hot and dense enough to be electrically conductive. Instead, the magnetic field might be offset from the center because the electrically conductive layer that is causing the magnetic field might be one of the layers above the core, in the planet's mantle.

"The electrical current system inside the planet, the circulation of ionized particles, is clearly not a simple, global thing," offered Ed Stone. "That may well have to do with the way the interior is differentiated." *Voyager* results and theoretical models appear to show that the mantle of Uranus is rich in water ice and other kinds of "volatile" molecules that started out as ices. At high enough pressures and temperatures, many of these ices, especially if they have hydrogen in their structure (such as H_2O), can become electrically conductive when vaporized and compressed. Some planetary scientists thought that the strange, offset, tilted magnetic field discovered by *Voyager* observations could be telling us that Uranus is really not like Jupiter and Saturn but is instead a completely different kind of giant planet. But it was hard to know, partly because the physical characteristics of Uranus are so different from the other giant planets.

Tilted on its side, not generating its own internal heat . . . "Clearly, something strange had happened to that planet," says Heidi Hammel. "It was hard to generalize about how Uranus-like planets could be so different from Jupiter and Saturn, because Uranus itself was just personally so screwed up."

During the days leading up to *Voyager 2*'s eventual closest approach, we all witnessed the five large Uranian moons go from mere specks of light to small disks to fully resolved worlds of their own. This was an especially exciting process for me. I hadn't been through the Jupiter and Saturn flybys in this way; I hadn't seen those dots slowly revealed as distinct worlds. Rather, like most people, the first time I saw Io or Europa or Titan revealed for what it truly is, was via one of the "greatest hits" close-approach photos that was published in the newspaper or shown on the evening news after each flyby. It was just—*bam!* Io has volcanoes! Or—*bam!* Mimas has a giant crater that makes it look like the Death Star! During *Voyager 2*'s Uranus approach, however, there were no such moments. Rather, the reality of these new worlds came into view slowly, over many weeks, with a grace and air of anticipation that more accurately reflected the gentle gravity assist that we were all actually going through, riding along with the spacecraft.

During those last ten hours, though, as *Voyager* plunged deeper into the gravity well of the seventh planet, we all experienced plenty of breathtaking moments. One by one the five large icy moons were revealed to us from *Voyager*'s high-resolution images, and a total of ten new, smaller moons would eventually be discovered lurking in the images. All the large moons are heavily cratered, attesting to their generally ancient ages. Oberon and Umbriel are the most cratered, suggesting that they have changed little during their more than 4-billion-year histories. It was a mystery to us why Umbriel is so dark compared to the other four large moons—indeed, it's still a mystery today. Perhaps its surface contained a higher fraction of carbon-bearing ices that have been darkened more over time by the constant irradiation from the solar wind. The other moons show more geologic diversity. Fractures of 1 to 3 miles deep as well as cliffs on Titania suggest a past active interior. Similarly large rifts on Ariel, as well as evidence for some sort of icy, perhaps cryovolcanic, flows, suggest past tectonic activity and internal heating on that world. But the most diversity, and the most vexing mysteries, came from *Voyager*'s high-resolution images of tiny Miranda, the innermost of the large Uranian moons.

Miranda is a small world, only about 300 miles across, comparable in size to Saturn's moon Enceladus. The expectation for such small worlds is that they are too small to have had active interior heating or large-scale geologic processes—too little heat inside, and what little heat there would have been at the beginning would have dissipated quickly. The reality for both has turned out to be dramatically different from the expectations. Miranda has been far and away the most geologically active moon in the Uranian system. The surface is a mishmash of heavily cratered, relatively bland terrain adjacent to patchwork patterns of bright and dark curving grooves and ridges that look like strange alien racetracks. One of the patchwork terrains has sharp corners shaped like a giant *V* or chevron. And scattered around the boundaries of some of these patchy terrains are enormous steep-sided cliffs of ice. In some places, if you were to fall off the edge, you would fall 6 to 10 *miles* before you'd hit the bottom. The 50,000-foot-tall ice cliffs of Miranda are on my bucket list of the most spectacular places in the solar system that I'd like to go photograph someday.

It was exhilarating watching these pictures coming in along with other members of the *Voyager* imaging team. Each photo would flash on the screen for a few minutes, and then the next one would replace it. I remember a stunned crowd of planetary geologists sitting around one of the worktables and watching the Miranda flyby image playbacks. A chorus of "Oooh!" "Ahhh!" "Wow!" and "What the heck is going on there?!" It was like watching a fireworks show that just kept getting better and better. People were giddy, and the geologists were deeply puzzled. "These objects are tiny—Miranda is only 1/100,000th the mass of the Earth. Yet this tiny world has giant ridge structures like racetracks curving across its surface," *Voyager* imaging team member Larry Soderblom said, clearly recalling being astounded by the diversity of the Uranian satellites.

Mission architect Charley Kohlhase once told me that in the beginning of the *Voyager* Project some people were worried about whether a lot of the moons that would be revealed would end up being sort of like the Earth's moon, heavily cratered and not particularly different from one another. "Would it be 'Once you've seen one moon, you've seen them all'?" he said they were asking. Happily, though, "that did not happen! And that was one of the great surprises of *Voyager*. There was no uninteresting moon. They were *all* interesting—from the volcanoes of Io, the cracks on Europa, the haze on Titan to the Death Star of Mimas. . . ."

Voyager imaging team member Rich Terrile was also both puzzled and delighted by how different the many worlds of the outer solar system turned out to be. "Before *Voyager*, we were kind of used to seeing a lot of craters, and a lot of 'boring' things," he told me. "Mars was only *just starting* to get interesting, with some evidence of streambeds and the like coming in from the *Viking* missions. But we really hadn't yet had that experience of seeing something for the first time and just *immediately* having your mind blown by something that flashes up on the screen. That had just not happened before. *Voyager* just turned the tides on everything. The outer solar system was so different than what we had expected. The joke was, the only thing you can *expect* from *Voyager* was to be surprised."

What *did* happen on Miranda? How could such dramatically different and bizarre kinds of geologic features coexist in such proximity? It was almost as if Miranda were a giant 3-D jigsaw puzzle that had been taken apart and then put back together, but with a bunch of the pieces twisted around or put back inside out. Indeed, some sort of massive but relatively gentle breakup of that moon, perhaps by a low-velocity giant impact or a tidal encounter with some other large moon, followed by reassembly, seems to be a leading hypothesis for what happened. Maybe Miranda was ripped apart and then poorly sewn back together long ago. Or maybe there are some kinds of geologic processes on small, icy, far outer solar system bodies that we simply do not yet understand. The feeling as *Voyager* sped on was partly exhilarating—no one ever expected to see such wonders—but also partly wistful and melancholy. The geology is so strange, so unexpected, and the encounter was so short . . . it could easily be many, many decades before we're able to go back and get a better look.

Still, no one could focus too much on the distant future, for there was science still to be done as the spacecraft glided past its closest approach and then through the shadow of Uranus to study the planet's atmosphere and rings. The rings had been discovered nine years earlier by a team of planetary astronomers led by the late Jim Elliot of MIT. Jim, his then grad student Edward "Ted" Dunham, and other colleagues were "occultation hunters," astronomers who could very accurately predict when a planet or moon or asteroid would pass in front of a bright star, allowing them to study the object's surface or atmosphere by the way that star's light was blocked, or occulted, as it passed behind. It was a neat trick, but it meant being nimble and flexible, because such occultations occur only rarely, and are visible only from certain very specific places on Earth—and almost never where a telescope has been built. So occultation hunters had to take their telescopes to the event, not the other way around. Jim and Ted and their colleagues built an occultation-chasing system that they could fly on NASA's Kuiper Airborne Observatory, a modified C-141A jet that flew missions in the stratosphere, above most of our atmosphere's clouds and water vapor. From there, they could chase occultations over a wide range of the Earth and be guaranteed good weather because the airplane flies above the clouds.

Ted was aiming to do his PhD dissertation research project on the composition of the atmosphere of Uranus by watching events like the March 10, 1977, passage of the bright star SAO 158687 behind the planet. As the starlight passed through the upper atmosphere before slipping completely behind, he'd be able to watch for telltale changes in the color and intensity of the star's light that would provide clues about the density, temperature, and composition of the gases that the light was passing through. Everything was set up perfectly for the experiment and as Uranus slowly moved closer to the star, they started recording data. At first, they thought that five little blips in the starlight that they saw well before the occultation with the planet were glitches in their setup, or maybe noise from the airplane or other systems. But then, after successfully recording the occultation, they saw the same five little blips just as far away from Uranus on the *other side* of the planet. It was as if the starlight had been blocked by five narrow rings around the planet. Wait—Uranus has rings! Subsequent observations revealed them to be much darker than Saturn's rings and confirmed the discovery of four more rings, making this what was then only the second known ring system in the solar system.

Motivated by this Earth-based discovery, *Voyager* imaging scientists were keen to find out if *all* the giant planets had ring systems, which led to the specific imaging sequences that would enable *Voyager 1* to discover the faint, dark rings of Jupiter in 1979. The best opportunity to study the rings of Uranus would come after *Voyager 2* passed the planet and was looking back toward the sun, using the same kind of light-scattering trick that had been used to study the Jovian rings. The planning paid off and *Voyager 2*'s images and its own stellar occultation data showed not just the nine previously known main rings around Uranus, but two more thick rings and a thin, dark, dusty sheet of ring material filling the dark bands in between (later, Hubble Space Telescope images would lead to the discovery of two more main rings, bringing the total rings around Uranus to thirteen). The Uranian rings are dark as charcoal and likely made of centimeter- to meter-sized blocks of icy, carbon-bearing materials that have been darkened by radiation from the solar wind and from the planet's magnetic field. Neptune, too, was later discovered to have dark rings like Uranus and Jupiter, further adding to the debate over the young versus old age of these kinds of ring systems, as compared to the brighter, "cleaner" ring system that the *Voyagers* studied around Saturn. One clue that suggested to *Voyager* scientists that the Uranian rings are young is the fact that they vary in width and thickness around their circumference, including some places where some of the thinner rings seem to disappear completely. While not definitive, this kind of variability suggests a young, evolving system that has not settled down into the kind of orderly, stable state that would be expected if they were ancient survivors from the formation of Uranus itself.

The year after *Voyager* flew by Uranus, I graduated from college and was accepted to graduate school in the Planetary Geosciences Program at the University of Hawaii in Honolulu. I was very lucky to get into grad school at all. Partly because I spent way too much time doing research rather than homework and studying, my grades at Caltech were awful, and I scored horribly on the physics part of the GRE test, so I'm sure I was a particularly weak applicant on paper. But luckily, I had spent a summer fellowship the year before working with colleagues from the University of Hawaii on some planetary astronomy research at Mauna Kea Observatory on the Big Island, and they thought I might be a bit more useful than my five pages of crappy grades and test scores suggested (the six other grad schools that I applied to didn't see it that way). Who am I to judge? I now often think to myself as I sit with other professors and read through applications for our own graduate program at Arizona State University. . . .

One of the opportunities that came up while I was in graduate school was a chance to finally try to do something useful scientifically with the *Voyager* images. NASA announced a program called the Uranus Data Analysis Program, which would enable researchers outside of the *Voyager* team to compete for funding to do new and different kinds of analyses of the data. I had no idea that grad students weren't allowed to submit such proposals, so I wrote up my project idea, estimated how much of my time I'd spend on it, put together a budget, and mailed in the proposal to NASA headquarters in Washington. About six months later, to my amazement, and to the surprise and consternation of my advisor and department administrators, the university got a letter back from NASA saying that they'd be happy to fund my research but wanting to know if there was a faculty member who could help to oversee the work. They were chuckling in DC (according to the NASA official who had helped to select my proposal), but they were discombobulated in Honolulu.

With my wrists duly slapped but the study green-lighted, I went to work on a project to map the different kinds of materials on the five major icy satellites of Uranus using not just the high-resolution black-and-white photos that *Voyager 2* had taken, but also the lower-resolution (more distant) color photos that were taken on approach. In retrospect, it was actually a really dopey and naïve kind of project, because the moons are very gray and show only small and not particularly diagnostic color variations on their surfaces. *Voyager* team members recognized this quickly, which is why (I believe) no one had yet gone off and worked on and published the project that I had proposed to do. Maybe NASA officials and my proposal reviewers realized this too, but since my proposed budget was so small, it was treated as a worthy student training project rather than the path to the next major discovery in planetary science. I had to learn, and in some cases create, image-processing software to work with the images, to do the color analysis, and to make the resulting maps. My results were essentially null—no major color variations and thus no new clues to the moons' compositions—but I found somewhere to publish my null result and in the process learned how to write a peer-reviewed scientific journal article. It felt good to have an official scientific connection to *Voyager,* albeit a weak one.

In the two decades since *Voyager 2* flew by Uranus, we've learned a lot more about the place from a variety of ground-based and space-based telescopic observations of the planet and its moons and rings. Perhaps the most striking discovery has been that we were fooled by the *Voyager 2* images into thinking that the planet's atmosphere is always boring. It turns out that *Voyager* happened to fly by at an anomalously boring time.

"We flew by at the peak of the summer solstice," Heidi Hammel says, "when the whole polar region was enveloped in haze. You couldn't see discrete cloud features in the *Voyager* images, and we didn't have an infrared camera that would let us see through that haze to the cloud layers below." In Hubble Space Telescope images as well as those from one of the giant 10-meter Keck telescopes on Mauna Kea taken since the *Voyager* flyby, many more visible bright and dark bands, clouds, and small storm systems have been tracked, revealing more complex and dynamic weather than originally thought. *Voyager 2* had flown by Uranus when the southern hemisphere was nearly continuously sunlit and the northern hemisphere was dark. Since then, the seasons have advanced through the southern fall/northern spring equinox (in 2007), and telescopic observations have revealed some significant seasonal changes. "It's a very different kind of planet now than the one *Voyager* saw, because of its dramatic seasonal changes," says Heidi Hammel. "Uranus has now traveled a quarter of the way around the sun since 1986, and now almost the entire planet is in sunlight, rather than just the south pole.

If we were to fly by Uranus with *Voyager* now, we would see a much more active planet than we did back then. It's kind of neat to watch those changes."

The *Voyager* flybys of Jupiter were followed up with a dedicated orbiter mission, *Galileo,* including an atmospheric entry probe that further revolutionized our understanding of the Jovian system. The *Voyager* flybys of Saturn were followed up with a dedicated orbiter mission, *Cassini,* which also carried a small probe, *Huygens,* that successfully descended through the clouds and hazes of Titan and landed on its surface. Similarly, many realize that the same path will have to be followed at Uranus (and Neptune), with a dedicated planetary orbiter being the next logical ramp-up in the exploration of that world. In the Interstellar Age, we know that to truly get to know a place, you've got to spend real time there, among the locals, learning their strange, alien ways.

7

Last of the Ice Giants

THE DISCOVERY OF Uranus in 1781 not only doubled the size of the then-known solar system, it also made English astronomer William Herschel a household name and a scientific pop star in his day. The idea of discovering an entirely new planet, and the resulting scientific immortality that would ensue, helped compel a revolution in eighteenth- and nineteenth-century telescope technology—an arms race, of sorts—among wealthy European gentlemen scholars and the patrons and potentates who supported them. Herschel had shown that cutting-edge technology, in this case building and having sole use of the largest telescope in the world, was a straightforward (though challenging and expensive) way to make new discoveries. Everyone wanted in, and new telescopes began popping up all over the continent. It was a growth industry.

Meanwhile, the mathematicians were in on the game too. Now that it was possible to accurately track the positions of Uranus and the other planets against the background stars with unprecedented precision, it was also possible to search for tiny deviations in their positions that could arise from the gravitational attraction that might be exerted by some new, as-yet-undiscovered planet. In some sense, just like the *Voyagers*, all of the planets, moons, asteroids, and comets in the solar system are constantly going through mini gravity-assist flybys with one another, slightly tweaking their positions relative to what would otherwise be the kinds of perfectly predictable orbital motions that Johannes Kepler and Isaac Newton had long ago described.

When my colleagues on the navigation team at JPL, for example, want to study a possible trajectory for a new space mission, they load their computers with the positions and masses of the sun, all the planets and their fifty or so large moons, and more than a half million asteroids, to make sure that every single possible "perturber" of the spacecraft is taken into consideration in their calculations. When astronomers and mathematicians like Edmond Halley and Pierre-Simon Laplace were working out the theory of motions of comets and asteroids, they were working on what physicists call the three-body problem, for example needing to account for the gravity and motions of the sun, Jupiter, and one of the Galilean satellites; or maybe the sun, Jupiter, and a newly discovered comet. Today's more sophisticated computer modeling of solar-system motions search for solutions to what is known as the n-body problem, whereby n is some very large number of objects. Everything exerts a force on everything else; we are all moved by the planets.

Two especially talented nineteenth-century mathematicians, John Couch Adams from England and Urbain Le Verrier from

France, were particularly well connected with the astronomy community and the best available data on the positions of Uranus. Without the use of computers or mechanical calculators of any kind, both men noted slight differences between the predicted and actual positions of Uranus, and both set about calculating the predicted position of a hypothetically more distant new planet whose gravity could be perturbing the orbit of Uranus, thereby explaining the differences. It was a classical three-body problem in which the bodies were the sun, Uranus, and some unknown but suspected massive new planet beyond Uranus. Although Adams and Le Verrier were working independently, and unbeknownst to each other, on the same problem, it was still a pitched scientific battle between the British and the French the likes of which as had occurred on any actual battlefield over the course of their histories, with nothing less than solar-system glory on the line. Adams asked his telescope colleagues at Cambridge to search for the putative new planet in a particular broad region of the sky. Le Verrier convinced colleagues at the Berlin Observatory in Germany to search a narrower predicted region (after failing to persuade his own institution, the Observatory of Paris, to spend more than a token effort on the search). Late in 1846, the French won the battle, as German astronomer Johann Galle found Le Verrier's "perturber" (on the first night that he looked for it), and confirmed its identity as the eighth planet.

The successful prediction and then discovery of a new planet was hailed as a triumph of modern physics and mathematics; according to French mathematician and politician François Arago, Le Verrier had discovered a planet "with the point of his pen." Le Verrier and Galle were jointly credited with the discovery of the new planet (even by the gracious Adams), and Le Verrier chose to name

it Neptune, after the Roman god of the sea. With the new planet orbiting at an average distance of thirty times the distance from the
Earth to the sun, once again, but not for the last time, the size of the
solar system was roughly doubled. While it was a modern triumph
of math and physics, there had also been a bit of luck involved—the
same kind of luck, in fact, that would enable *Voyager*'s Grand Tour
mission to take advantage of the discovery of this new planet some
130 years later. Both Adams and Le Verrier were searching for per-
turbations in the orbit of Uranus using measurements of the position of Uranus taken between 1846 and the time of its discovery,
back in 1781. During that specific time period, Uranus *happened* to
have passed Neptune in its orbit (around 1821, actually, before Neptune was discovered), and thus Neptune *happened* to have had its
maximum influence on the orbit of Uranus right when the telescopic
and computational "technology" of the day would allow such tiny
tweaks to be recognized. Such an alignment hadn't occurred since
the year 1650, when neither planet was known, and wouldn't occur
again until the year 1993—by which time technology had advanced
so far that we had actually visited both worlds with robotic spacecraft. That same time interval of about 175 years between alignments of Uranus and Neptune is what Gary Flandro and others
recognized, in the 1960s, sets the timing of potential spacecraft
Grand Tour trajectories to all four giant planets.

OHANA

As *Voyager 2* sailed on toward Neptune after its 1986 encounter
with Uranus, my own trajectory moved me distinctly westward, to

find my way as a professional planetary scientist via a graduate education in Hawaii. The late 1980s and early 1990s were a particularly challenging time for NASA's exploration of the solar system. The *Viking* missions to Mars, highly successful orbiters and landers, were over, and the future of new "flagship"-class missions was in doubt because of government belt-tightening. The loss of the Space Shuttle *Challenger* and her crew in 1986, just a few days after *Voyager 2's* flyby of Uranus, threw NASA's human exploration program into a state of disarray, and along with it many of the robotic planetary science missions, such as the *Magellan* Venus orbiter and the *Galileo* Jupiter orbiter, that had been planning to use the Shuttle as a launch vehicle. The idea of intermediate- and smaller-class, "better, faster, cheaper" planetary space missions hadn't been invented yet. In short, it was a great time to be focusing on the use of "sure bet" facilities, such as high-powered ground-based telescopes, to try to push the envelope in planetary science. So that's what I did, and there was no better place than Hawaii, as I had learned during my summer spent at Mauna Kea Observatories between my junior and senior years in college.

In yet another example of the lingering power of astrology—the "undeniable" ability of the positions of the planets to influence people's lives—the quirk of the timing of my birth in 1965, combined with the typical track of elementary, secondary, and college education in the United States, put me in my first years of graduate school in the mid- to late-1980s, right when, coincidentally, the Earth and Mars happened to be going through a series of close passes every few years that provided outstanding opportunities for the best telescopic observations yet made of the Red Planet. My PhD research group was working on just that, and so it became incredibly

convenient, as well as fun and interesting, to consider that as a potential thesis topic. If I'd been born five or ten years earlier, maybe I
would have pursued some other planet, or the more "pure" astronomy of stars and galaxies. If I'd been born five or ten years later,
maybe I would have skipped learning how to research with telescopes entirely and tried to go straight into space missions, like
many modern planetary science students do today. But no, my birth
in July of 1965 set me on an astrological collision course with Mars
via telescope in the late 1980s, and then a lifelong involvement in
Mars via spacecraft thereafter. Is it a coincidence that the world's
first spacecraft flyby mission past Mars, *Mariner 4*, *also* occurred in
July of 1965? Well, yes, in fact it is.

Though I was primarily studying Mars in grad school, I couldn't
let go of *Voyager*. I wasn't at Caltech anymore and no longer had a
direct connection to the mission via my mentor Ed Danielson or
any of my other former professors involved in the mission, like Andy
Ingersoll. Still, I had to figure out a way. . . .

Planetary science is a rather small, close-knit field. When I was
in graduate school, there were maybe five hundred or so professional
planetary scientists, including grad students, in the field. Nowadays
the number is something like three or four times that, but it's still
small enough that most people know one another, or at least know *of*
one another. There's a bit of a family feel to the endeavor, with most
of the community getting together sort of "over the holidays"–style
(at one or the other of the major conferences held during the course
of the year), or for special occasions, like a spacecraft launch, flyby,
or landing.

After moving to Hawaii from Caltech, I was doubly fortunate to
find *two* amazing families there in the islands—one made up of

close friends and mentors who helped me in my work world at the university, and the other made up of close friends and mentors who helped me after hours when I learned how to paddle as part of a Hawaiian outrigger canoe club. Paddling with my brothers and sisters in the waves off Waikiki Beach, learning about ancient Polynesian navigation and other local traditions, and kicking back afterward to soak in some "island style" music and food taught me not just a new word—*ohana*, Hawaiian for "family"—but the inner spirit behind the word as well.

If any one person was the embodiment of *ohana*, it was Fraser Fanale, a professor (now retired) of planetary science who specialized in thinking about the history of water and other "volatile" molecules in the solar system—on Mars, on the satellites of Jupiter . . . anywhere. He would talk to anyone, at any time, about volatiles, where to find them, how they move around, and what they tell us about how planets, moons, asteroids, and their environments evolve over time. Fraser was one of the most disorganized, but kindhearted, souls I have ever met. He defined the stereotype of the absent-minded professor. In the days before cars had those high-tech key fobs that let you flash the lights or honk the horn, I'd run into Fraser at conferences in the parking lot to the hotel, wandering around trying to find his rental car, which he'd misplaced. He showed up to give a talk once with a stack of viewgraphs—handwritten overhead projector notes—and proceeded to drop and scatter them while being introduced. Undaunted, he delivered his talk, legibly but out of order, from the randomly reassembled stack. Sometimes it was almost comical, but—my goodness!—the man had an impressive and intuitive grasp of what is going on out there in the solar system.

Fraser hadn't been directly involved with *Voyager,* but he was a member of the *Galileo* Jupiter orbiter mission team, helping plan for observations of volatiles on the Galilean satellites. One of the ways *ohana* manifests itself in planetary science is that, almost always, members of a spacecraft team who are preparing for some publicity associated with an upcoming event, like a launch or, say, a flyby of Neptune, will often invite members of *other* mission teams, and their families, to participate in the event. So, early in the summer of 1989, Fraser found himself invited to attend the events at JPL surrounding the historic *Voyager 2* flyby of Neptune in August. Fraser wasn't a big fan of traveling, and by this time the first prototype versions of something called the Internet were appearing on college campuses and government labs, allowing colleagues to exchange "electronic mail" messages and to send digital photos back and forth over a strange new interconnected series of dedicated communication lines called the World Wide Web. Fraser was going to sit it out, watching the Neptune images come in over the web. But his invitation and badge were transferrable, and he knew about my work with the *Voyager* images of the Uranian moons. Would I like to go? he asked.

Dude.

THE OTHER BLUE PLANET

Voyager 2 passed Uranus in 1986 at exactly the right place and time to slingshot onward to where Neptune *would be* in August of 1989. The precision with which the JPL navigation team guides spacecraft like the *Voyagers*, and others, still astounds me. Three separate "needles" had to be threaded—at Jupiter, Saturn, and Uranus—precisely, to get the spacecraft where it needed to be for our first and only en-counter with Neptune. The physics behind the feat had mostly been worked out in the seventeenth century by Isaac Newton. But New-ton could take us to the planets only in a hypothetical, theoretical way. "It's not rocket science," joked one of my JPL friends when I tried to gush over the remarkable and historic exploration achieve-ments that she and the extended *Voyager* team had helped pull off. Oh, but it is. Working out the mission on paper is one thing (and an impressive thing at that). But to actually *get there* required techno-logic feats and innovations not possible until the late twentieth century, not to mention the largesse of the taxpayers of an entire nation to help foot the bill.

Just like in the lead-up to the Uranus flyby, improvements were made remotely to the spacecraft as well as to the ground stations that would be needed for the encounter. *Voyager*'s thrusters were throttled down even more to allow the spacecraft to slew even more slowly and gracefully—essential for the dimmer level of sunlight at Neptune compared to Uranus. Improvements were made to the image motion compensation software and to the camera software to enable it to compress images better and to take longer-exposure images without smearing. And improvements in NASA's Deep Space Network were needed to reliably detect the extremely faint radio signals from the spacecraft. *Voyager 2* had traveled so far from home that signals from Earth to and from the spacecraft as it neared Neptune were now taking more than four hours, each way, traveling at the speed of light. Larger radio antennas (now nearly 230 feet across), and new receiving stations in New Mexico and Japan, helped guarantee high-quality communications with the spacecraft.

"We had a huge number of antennas online," Charley Kohlhase recalls, "and that allowed us to get respectable data rates—tens of kilobits per second!" While still much slower than good old dial-up computer modem speeds, this was not a bad data rate for an interplanetary Internet connection that had to span a distance of 2.7 billion miles. *Voyager* science sequence coordinator Randii Wessen, thinking back to the impressive collection of nearly thirty giant radio telescopes from the West Coast of the United States to Australia to Japan that were all linked together or "arrayed" to pick up the spacecraft's faint signal at Neptune, told me, "We joked that we were listening to *Voyager 2* with the entire Pacific Basin!"

There was also extensive speculation among planetary scientists about what *Voyager* would find at Neptune. Sunlight at Neptune is a mere *3 percent* as intense as sunlight at Jupiter, so some colleagues figured that Neptune would have a relatively bland and featureless atmosphere, like Uranus, because of the lack of solar energy to power the kinds of intense storms seen in the gas giants closer to the sun. Others speculated that Neptune could have substantial reserves of internal energy, like Jupiter and Saturn, which could power significant atmospheric activity. If the pattern held, Neptune would have a powerful magnetic field, and the field would be more like Jupiter's and Saturn's because Neptune's spin axis is also hardly tilted at all. The radiation from that magnetic field, along with the solar wind, would probably darken the rings (rings that were discovered earlier in the decade from ground-based telescopic occultations, the same method that had led to the discovery of the Uranian rings) and any small satellites in the vicinity of the planet. Many of the predictions were fairly safe to make, as they reflected a newfound understanding—made possible by the *Voyagers*—of some of the basic properties and processes that occur on, in, and around gas giant planets.

More speculative, however, were the predictions of what Neptune's large moon Triton (not to be confused with Saturn's moon Titan . . .) would be like. Triton, which was discovered in 1846, just a few weeks after Neptune itself was discovered, is an oddball because it is one of the largest moons in the solar system (a fact known well before the *Voyager* flyby) and it orbits Neptune *backward* relative to Neptune's spin direction, and at about a 25-degree tilt relative to Neptune's equator. Triton is the only large, planet-sized moon in the solar system that orbits its planet backward relative to the spin direction of its parent. Partly because of this, astronomers believe that Triton was formed elsewhere and was then somehow "captured" by Neptune, as no one's computer model or thought experiment can come up with a good way to explain how it could have formed in a backward, tilted orbit. But *where* would it have been formed? How did it get captured, and why backward? Would it be *anything* like any of the other big icy moons seen previously by *Voyager*? Fortunately, since Triton would be *Voyager 2*'s last port of call on its Grand Tour, no one had to worry about exactly where the spacecraft needed to be "next." So mission planners could target *Voyager*'s aim point very close to and through the shadow of Neptune (passing high enough above the planet's north polar atmosphere to avoid any potential drag on the spacecraft or electrical arcing from friction with upper-atmosphere ions), so that the planet's gravity would divert the spacecraft toward a close flyby of enigmatic Triton about five hours later. Charley Kohlhase has estimated that the accuracy needed to deliver *Voyager 2* to its aim point above Neptune to within about 62 miles after traveling on an arc more than 4.4 billion miles long is roughly equivalent to a golfer sinking a putt from Washington, DC, to Phoenix, Arizona. Although he did acknowledge that a few "fine adjustments" (a.k.a. cheats) might be needed along the way. Amazingly, after the flyby, the actual navigation errors were found to be roughly ten times smaller than the requirement!

Voyager science team members Candy Hansen and Torrence Johnson recall some particularly anxious moments as they were trying to plan the photographs of Triton. No one had ever seen the place up close, of course, and so there was significant uncertainty about how best to take the photos. From Earth-based telescopic observations, Triton was just a point of light near Neptune. "By Saturn we felt more comfortable making exposure-time estimates from ground-based observations, but the scariest was at Neptune with Triton," recalled Johnson. It reflected a certain overall amount of sunlight that could be very accurately measured, but was it as bright as observed because it was very reflective (maybe icy) and small, or less reflective (like rock or soot) and large? Team members took some images of Triton with the *Voyager* cameras during the cruise from Uranus to Neptune to try to get more information. It was still just a point of light, but by photographing it from different angles compared to Earth-based telescopes, they hoped to gain some insight on what to choose for the exposure times. This kind of building up a set of images from different angles is part of characterizing what is esoterically called the *phase function* of a planet or moon or asteroid, and it is a common way for astronomers to obtain information remotely on what kind of materials—for example, icy or rocky or metallic—a surface is made of.

Even while still very far away, as *Voyager 2* approached the

Neptune system, the phase function data were suggesting that Triton was smaller and brighter (with a very reflective, more icy, surface) than their initial estimates. "We switched to shorter exposures on this basis, just weeks before encounter," Johnson said. "We were changing exposure times in every iteration of the sequence," Candy Hansen added, "until finally the sequence was on the spacecraft and we just couldn't change it anymore. Fortunately by then we had nailed it!" Time to bake more cookies!

The summer months leading up to the *Voyager 2* flyby of Neptune in 1989 were filled with all kinds of anticipation and excitement as the planet slowly grew in the images from a point of light to a resolvable orb. I was following along remotely from Honolulu for most of the approach, using my new e-mail account to receive occasional images from Ed Danielson or other colleagues on the imaging team. Neptune was clearly going to be different from Uranus. First of all, the color of the planet was different—Neptune's strikingly blue color was apparent even from images taken months before the flyby. At Uranus, the planet's aquamarine color was a telltale sign of the presence of methane, which absorbs red light scattered through the atmosphere. In advance of the flyby, it was theorized that Neptune's azure hue was also due to methane, but at a colder temperature and higher pressure than at Uranus, causing it to absorb not just the red colors in the sunlight but some of the green colors as well. *Voyager*'s instruments would eventually provide flyby data that would prove this to be the case, but in the meantime, we were all just marveling at the fact that our solar system had a second blue planet.

And it wasn't just the color of Neptune that was causing surprise and delight. Folks like Andy Ingersoll and others on the

imaging team had planned another approach movie for this flyby—
a series of snapshots taken at least every day, and sometimes many
times per day, during the long approach to the planet. Even from
more than 60 million miles away in June of 1989, more than two
months before the flyby, cloud features could be clearly seen and
tracked in Neptune's atmosphere. Darker and lighter bands could
be seen at different latitudes on the disk of the planet—not as colorful and artistic as the belts and zones of Saturn or, especially, Jupiter,
but more dramatic than the relatively bland atmosphere of Uranus.
Candy Hansen and brand-new *Voyager* imaging team member
Heidi Hammel spent a lot of late-night time doing science at the JPL
browse workstations. They were among the first to notice a large,
dark, oval-shaped feature in Neptune's southern hemisphere. It had
a similar shape to Jupiter's Great Red Spot, and so it was quickly
dubbed the Great Dark Spot.

Soon, small white features began to be resolved in the images,
spinning around the edges of the Great Dark Spot. Those features
eventually came into view as clouds, cementing the idea of the Great
Dark Spot being a giant (Earth-sized!) storm system, probably
much like Jupiter's Great Red Spot. As the spacecraft got even closer,
smaller white clouds and other dark spots came into view, and calculations by Heidi and Candy and others on the team showed that
some of them appeared to scoot around the planet at much faster
speeds than the Great Dark Spot—indeed, Neptune's features have
some of the fastest wind speeds ever clocked in the solar system (up
to 1,300 miles per hour!). There was clearly a lot going on.

"Every day when we came in, it was like the veil of mystery
around Neptune was getting thinner and thinner and thinner,"
Heidi says, reflecting back on that pre-encounter approach phase

when Neptune went from the small, fuzzy blob that she knew as a telescopic observer to the richly detailed world that *Voyager* observations would reveal it to be. "It was just incredible," she mused. "Every time a new series of images would come down, there was sort of a gasp, and a moment of excitation, and we would say, 'Oh wow— that's amazing. . . .' The giant fuzz blobs that I'd been tracking from Mauna Kea turned out to be just one feature out of many features." In what she amusingly describes as her "great tragedy of the Neptune encounter," she recalled how she was sent back to Mauna Kea to take photos of Neptune at exactly the same time as *Voyager* during the closest approach. "'We're going to need ground truth,'" she recalls Brad Smith telling her, "so we know how to compare our spacecraft data with previous and future ground-based data." She remembers saying, "Yeah, that makes a lot of sense, Brad," to which he replied, "And you're the world's expert. . . ." So she spent the entire flyby on the summit of Mauna Kea, hungrily devouring fax messages from imaging team colleague Andy Ingersoll telling her what they were seeing in the *Voyager* images back at JPL. "I missed the whole thing," she lamented. "But I guess that's what you do for science."

While dazzling, the higher level of atmospheric activity on Neptune compared to Uranus was also extremely puzzling. Neptune gets 40 percent less solar heating than Uranus, and so if sunlight is what is driving the energy of these giant-planet atmospheres, then Neptune should be even less active than the relatively bland Uranus. Indeed, the pattern observed by *Voyager* of the decreasing numbers of clouds, belts, and storms as the mission traveled from Jupiter, out to Saturn, then farther out to Uranus, was consistent with this idea of solar energy mostly powering the weather on these worlds. But Neptune proved that simple explanation wrong, dramatically.

One of the things that *Voyager* was able to measure during the flybys was the total amount of heat energy coming out of each giant planet. If they were in balance with the heat energy provided by the sun, then the total amount of solar energy going in would equal the total amount of thermal (heat) energy coming out. This kind of balance is a fundamental feature of the terrestrial planets Venus, Earth, and Mars, where the temperatures of the surfaces and atmospheres of those worlds are very different but, ultimately, driven by solar heating. At Jupiter, however, *Voyager* measured almost *twice* as much heat energy coming out of the planet compared to the solar energy going in. The same was true at Saturn. Clearly, for those giant planets, there must be additional internal heat sources that contribute to the energy in their atmospheres. Planetary scientists speculate that the extra heat could be the result of the enormous amount of gravitational energy stored in the high-pressure, high-temperature deep interiors of these planets, or possibly from heat released by the decay of radioactive elements within the rocky cores thought to exist deep inside them, or even from heat released during chemical reactions within each planet as materials change from one phase (such as ice) to another (such as vapor) with rising pressure and temperature.

At Uranus, however, the amount of heat energy coming out was essentially *equal* to the amount of solar energy going in. This makes Uranus fundamentally different from Jupiter and Saturn. Aha! figured the science community: smaller gas giant planets don't have internal heat sources like the much larger Jupiter and Saturn. But then, *Voyager 2* measured *three times* as much internal heat energy coming out of Neptune compared to the solar energy going in.

Another monkey wrench thrown into the gears of understanding—to the delight of many of my colleagues. New discoveries lurk within the details of the unexpected.

Just like for Uranus, telescopic observations from the ground and from space since the *Voyager 2* flyby have revealed major changes in the atmosphere of Neptune over time. First, the Great Dark Spot in the southern hemisphere disappeared. Then a different Great Dark Spot formed in the northern hemisphere, along with a second northern dark spot. White clouds and smaller spots fade in and out, and the belts at different latitudes brighten and darken over time.

"We seem to see a new Great Dark Spot form and then dissipate about every five years, but we don't know why," Heidi Hammel says. "But we aren't able to look very often with high-resolution tools like HST or Keck. We see one, then we don't, but we don't know what happens in between. We really need more continuous coverage of the planet's weather to be able to track the features and figure this place out." It is a dynamic atmosphere, and still largely mysterious, as we've been studying the place at high resolution for only a small fraction of Neptune's 165-Earth-year trip around the sun (indeed, we've only known about the place at all for just one Neptune year). And even today, the reasons for Neptune's strong internal heating, and for the lack of strong internal heating on Uranus, are not clearly understood.

Voyager's measurements of the overall chemistry and interior structure of Uranus and Neptune have led to a transformation in the way we view these giant planets as compared to their larger Jovian-class cousins. *Voyager* helped us look inside these worlds, revealing that while the outer layers and the visible "surfaces" of all four of the giant planets are made of clouds and gases, deep in their interiors they differ from one another in ways that are not obvious from our vantage point on Earth. As Jupiter and Saturn were forming some 4.5 billion years ago, they captured huge amounts of hydrogen and helium in the cloud of gas and dust (called the solar nebula) from which our sun and the rest of our solar system was then forming. This gaseous envelope surrounds and dominates the (relatively) tiny, Earth-sized, rocky/metal cores of Jupiter and Saturn, meaning that they have essentially the same hydrogen-rich composition as the sun. Deep inside, that hydrogen acts like a metal at super-high pressures and temperatures, conducting electricity and powering those planets' giant magnetic fields. Jupiter and Saturn are, truly, gas giants.

In contrast, while Uranus and Neptune were forming early in the history of the solar system, there was not as much gas available farther away from the sun. At the colder temperatures of the far outer solar system, a lot more ice was condensing out of the solar nebula than in the warmer regions closer to the sun. The end result appears to have been that the interiors of Uranus and Neptune are made of a relatively larger fraction of vaporized ices (like water ice, methane ice, ammonia ice, and other volatiles) than the interiors of Jupiter and Saturn are. These smaller worlds also each have an Earth-sized rocky/metallic core, but it is surrounded by a deep mantle of high-pressure, high-temperature vaporized ices, which is then surrounded by a relatively thinner, though still hydrogen-rich, gassy atmosphere. Uranus and Neptune aren't really gas giants, then: they are ice giants, dominated by a larger fraction of initially icy materials than their classical gas-giant cousins Jupiter and Saturn. *Voyager* observations had revealed an entirely new and unanticipated class of planet.

"I think the idea of ice giants as distinct kinds of planets from gas giants was developed during that year leading up to the Neptune encounter," Heidi Hammel recalled. Neptune was different. It has a perfectly normal tilt, and an internal heat source like Jupiter and Saturn. "But as *Voyager* drew closer, it became clear that even though it is a giant, this planet is *not like* Jupiter or Saturn. It didn't have the swirling cloud patterns that Jupiter and Saturn did. Even though it had this big Great Dark Spot, the more we saw of it, the less it looked like the Great Red Spot. It wasn't stable and round, but had a weird oval shape. It had all these bright companion-cloud features that shifted all around, sometimes under it, sometimes across it.... Even the smaller features were just very different than the small features on Jupiter and Saturn. The way the clouds were forming was entirely different. The closer you got, the more they resolved into tiny spots, like tiny clusters of connected thunderheads." The idea of Uranus and Neptune as fundamentally different beasts in the planetary zoo was part of an evolving understanding rather than an instant realization.

The prediction that Ed Stone and others had made about Neptune's magnetic field came true: it is strong and behaves in some ways like the fields around Jupiter and Saturn. But like the field inside Uranus, it is offset relative to the center of the planet, and tilted relative to Neptune's spin axis. Maybe, then, the strange tilt and off-set of the field at Uranus is not a feature of a strangely tilted planet but is instead a feature of any planet with an electrically conductive middle layer or mantle of high-pressure vaporized ices. Maybe all ice giant planets have tilted, offset magnetic fields. Certainly all the ones in our solar system do.

"Uranus and Neptune are not just like Jupiter and Saturn, except blue," said Heidi Hammel. "The processes going on there are fundamentally different."

Voyager 2 made other exciting discoveries in the Neptune system. Based on the earlier Earth-based telescopic discovery of at least partial rings around the planet, *Voyager* imaging team planners were able to design special imaging observations that took advantage of being able to look back toward the sun at the areas where the rings should be, enhancing the ability to see super-fine particles in the rings and to tell if they were complete rings or just partial arcs of material orbiting the planet (a question that was driving the mathematicians crazy, because such structures were predicted to spread out and turn into full rings in only a few years). The imaging worked beautifully and revealed a curiously stunning system of at least five separate rings around Neptune that are complete but *clumpy*, with the thickest, coarsest clumps corresponding to the ring arcs that had been seen from Earth, and the thinnest parts consisting of dark, fine-grained, very dusty materials that are too faint to have been seen from Earth.

Voyager 2 flew right through the gap between the two outermost rings but was still impacted by hundreds of tiny dust particles per second for several minutes. Luckily, it emerged unscathed, perhaps because the dusty ring particles are so small—only about 1/100th the width of a human hair. It's still not known what makes the rings clumpy, though planetary scientists suspect that Neptune's ring clumps may be getting "shepherded" in their orbits, like some of the thin rings of Saturn, by little moons that were too small and faint for *Voyager* to see. The rings were eventually named after early astronomers who were instrumental in the initial discovery and characterization of Neptune, including Le Verrier, Galle, and Adams, and the thickest clumps in the outermost Adams ring have been named Liberté, Égalité, and Fraternité, in honor of the fact that it was France that in 1846 took victory in the race to discover Neptune.

I spent the week surrounding the *Voyager* Neptune flyby back in Pasadena, having been granted a magic access badge through Fraser Fanale's invitation, and maybe also owing to the fact that enough of the team knew and remembered me from the Uranus flyby that they figured I wouldn't cause too much trouble. And this time, as a graduate student in the field, I might even be useful. Upon setting foot once again in those rooms where, brand-new, the images were streaming in from the farthest reaches of our solar system, I happily filled the role of gofer and errand boy for the real members of the imaging team. I was dutifully taking lunch orders, making copies, and otherwise trying to stay out of the way but still soak it all up, as if through some sort of academic osmosis. This would be *Voyager*'s last stop on the way to the stars.

THE NON-PLANETARY SOCIETY

By Neptune, *Voyager* team members like Rich Terrile and others had become very good at hunting for small new moons or strange patterns in the rings around giant planets, with the team having so far discovered twenty moons around Jupiter, Saturn, and Uranus, and thus having increased the number of then-known moons in the entire solar system by nearly 70 percent.

"That was a lot of fun," recalls Rich. "As an astronomer, you're trained in trying to pull some useful signal out of noisy data, and hunting for new small moons was exactly that kind of problem." One of his *Voyager* team colleagues challenged him at one point, stating flatly that "it's no big deal to find these things; they're just in the data." So Rich challenged that person back, giving them some images with recently found moons to look through as a test—which they promptly failed. The reason why imaging team members like Rich Terrile could spot these subtle little moons so quickly was because they'd had years and years of experience studying all the quirks and characteristics of stars, cosmic-ray hits, and camera artifacts in the *Voyager* images, and they could tell when a new little dot in an image wasn't any of those—especially if it was moving from image to image.

True to form, as the final pre-flyby *Voyager* approach images were streaming in, Rich and other team members were busily starting the process of discovering six small new moons of Neptune that hadn't been seen or confirmed from previous telescopic observations. Candy Hansen recalls the lighthearted teasing that Rich got from the team about his prowess for discovering new moons or features in *Voyager* images. "Like the time we told Rich that he couldn't come drinking with us at the Loch Ness Monster [a bar nearby the lab] until he discovered something," she recalled, laughing. "And then he showed up at that seedy, seedy bar with a hard copy of a *Voyager* photo showing an elliptical ring!" Rich earned his drink that night, and many others.

"That really happened," Rich confirms. "I was perfectly OK to stay at work and find more things, though. I was on a roll!" Combined with the giant moon Triton (about the size of Europa or our own moon), and the smaller, elliptically orbiting moon Nereid, these new moons took Neptune's then-known moon count up to eight. Since *Voyager*, six more smaller, fainter moons have been discovered by more sensitive ground-based and space-based telescopes.

If the moons of Neptune were to be the last planetary hurrah for *Voyager*, it would head for the stars with fanfare. After swinging just 3,000 miles above the cloud tops of Neptune (closer than to any planet that the spacecraft visited since leaving Earth), *Voyager 2*'s trajectory would take it past one final, glorious, unknown destination: Triton. The moon was discovered shortly after Neptune itself in 1846 (because it is so big and so bright), and *Voyager* imaging team members figured it would be an oddball of some kind because of its unusual, backward orbit. The spacecraft would fly within 25,000 miles of Triton, taking pictures of surface features as small as 5 to 10 miles across, and no one really knew what to expect. Because it is so bright and so far from the sun, Triton's is among the coldest natural surfaces in the solar system, with an average temperature only about 38 degrees above absolute zero (or an incomprehensible −391°F). Triton's brightness suggested that there would be relatively clean ice on the surface, perhaps even including exotic, low-temperature ices other than water ice. And its strange backward orbit suggested that it may have been through some sort of planetary-scale trauma, such as being captured by Neptune, or had its course changed by some sort of giant impact. It was a great way to end the surface-imaging phase of a great mission—with an encounter that would be surprising no matter what was revealed.

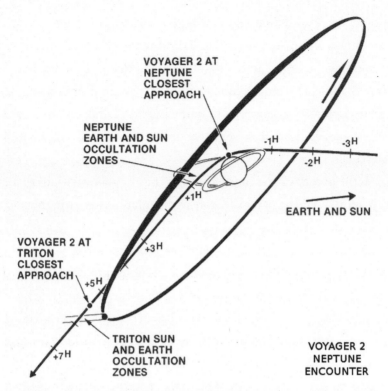

**VOYAGER 2 AT
NEPTUNE
CLOSEST
APPROACH**

**NEPTUNE
EARTH AND SUN
OCCULTATION
ZONES**

-₁H   -₃H

-₂H

+₁H

EARTH AND SUN

**VOYAGER 2 AT
TRITON
CLOSEST
APPROACH**

+₃H

+₅H

**TRITON SUN
AND EARTH
OCCULTATION
ZONES**

**VOYAGER 2
NEPTUNE
ENCOUNTER**

+₇H

Last Port of Call. *Voyager 2* flyby trajectory past Neptune. *(NASA/JPL)*

About five hours after closest approach to Neptune, *Voyager 2* flew past Triton. Several days later, I remember being in the JPL workroom with imaging team member Larry Soderblom, looking over the first high-resolution images of Triton that had come in. Larry is a friendly, outgoing, sometimes mischievous, and highly respected member of the planetary science community who works at the US Geological Survey's Astrogeology Science Center in Flagstaff, Arizona (I've always been confused about the name of that group, as *astrogeology* technically means "the geology of stars," but stars don't have geology—planets and moons and asteroids and comets do. But then again, "astronauts" don't travel to the stars … at least, not yet …). Larry was one of the *Voyager* imaging team members who would occasionally give me a nod and a wink and beckon me away from some dark, out-of-the-way corner to come sit at the big table and look at some images. How could I resist?

Luckily, I didn't feel too stupid, because what bizarre images they were! Larry was as mystified as I was. Instead of a surface covered with classic impact craters, cracks and ridges, or other typical features like those that had been seen on many icy moons before, Triton was determined to be different. The part of Triton's southern hemisphere that was sunlit during *Voyager*'s flyby was split into two very weird kinds of terrains: a darker one consisting of pits and dimples and ridges reminiscent of the skin of a cantaloupe, and a brighter one consisting of smoother plains materials interspersed with terraced depressions that looked like frozen lakes. A translucent, reddish layer of nitrogen ice in some places, and nitrogen snow or frost in others, appeared to drape much of the terrain. That, and the relative lack of impact craters, suggested that the surface was geologically very young. It was as strange and unexpected a place as *Voyager* had ever revealed, and I remember Larry laughing to himself more than once. "Isn't that just beautiful?" he would ask rhetorically.

To top it off, *Voyager*'s ultraviolet spectrometer team had recently discovered a surprise—Triton has a very thin atmosphere (less than 0.001 percent of Earth's pressure) made mostly of nitrogen and methane. Larry was looking for some evidence of the interaction of that thin atmosphere with the surface. He and others had already noticed dark wind streaks near the south pole—places where that thin atmosphere appeared to be moving sediments across the surface. Using software that he and his USGS colleagues had developed, Larry was making short time-lapse movies of Triton's surface as *Voyager* sped past, looking for evidence of any changes that the wind might be actively making. That's when they noticed something remarkable: dark plumes, rising up more than 5 miles above the brighter surface of Triton and then spreading out more than 60 miles downwind. Four of them had been caught in the act of erupting as *Voyager 2* flew past, and photographing them at different angles during the flyby is what allowed Larry and colleagues to view them in stereo and determine their heights.

"I was analyzing newly received Triton images with longtime USGS Flagstaff friend and colleague Tammy Becker," Larry recalls, thinking back to that wonderful moment of discovery. "We were building a new map of Triton's surface, pasting the overlapping *Voyager* images together into a mosaic. Because the flyby images were all taken from different angles as we flew past, we had to try to paste the images onto a spherical model of Triton's surface so they could be aligned into a global map. But some dark and bright streaks seen in two particular overlapping images just would not line up. We puzzled a bit and the reason soon became clear—the streaks were not *on* the surface but were *above* it! We put those two images together into a stereo viewer, grabbed our red-blue glasses, and Triton's plumes popped out of the surface and into full view!"

Active geysers on Triton! Larry and the team came up with a model to explain what they were seeing: sunlight warms the bottom of a relatively transparent 3-to-5-foot-thick layer of seasonal surface nitrogen ice, causing it to sublime (evaporation of ice directly from a solid to a gas) and collect under pressure under the ice. When the ice cracks somewhere, the nitrogen gas is explosively released, carrying dark dust and mineral grains along with it up to high altitudes under the low gravity and atmospheric pressure. It was electrifying to be around the team that was discovering and trying to explain, as the discovery was made, what was then only the third known place in the solar system with active eruptions of some sort (the others at the time were the Earth and the *Voyager*-discovered plumes of Io). I've since worked with Larry on the Mars rover missions as well as being the PhD dissertation advisor of his (equally mischievous) son, Jason. The baton is passed along.

We haven't been back to Triton since *Voyager 2* zipped past, but we have made some progress in trying to figure out where Triton may have come from. The discovery of Pluto, in 1930, was among the earliest pieces of evidence that astronomers used to argue for the existence of a large disk of similar small bodies extending well beyond Neptune's orbit. Among these visionaries was one of the fathers of modern planetary science, the Dutch-born American astronomer Gerard P. Kuiper. In the early 1990s, planetary astronomers began discovering the first members of that population beyond Pluto—the Non-Planetary Society, you might say. Today, taking advantage of substantial improvements in telescopes and camera detectors over the past few decades, more than 1,200 of these Kuiper Belt Objects, or KBOs, as they are now known, have been discovered. Many of them, like Pluto, are in an orbital resonance dance with Neptune that always keeps them far away from that giant planet's gravity (Neptune orbits the sun exactly three times for every two times Pluto orbits the sun). Over the 4.5-billion-year history of the solar system, it is hypothesized that some fraction of KBOs that were in unfortunate orbits bringing them too close to Neptune either crashed into the planet or were slingshot out of the solar system or even into the sun. But, just possibly, one of them—Triton—survived that close encounter and was captured by Neptune's gravity.

Voyager's flyby of Triton, then, may have been humanity's first encounter with a Kuiper Belt Object, a small planetlike body that originally formed in the cold reaches of the outer solar system, and which may provide a glimpse into the ices and rock that formed the original cores of all the giant planets. To illustrate how rare and precious the Triton flyby was, we would have to wait twenty-six years for our second encounter with a KBO, when the NASA *New Horizons* spacecraft flies by Pluto in July of 2015. After more than nine years of traveling through space as the fastest mission ever launched, *New Horizons* will have to do all its best science, and take all its best images, within about a thirty-minute period around closest approach to Pluto (no pressure on that team, eh?).

There's a lot of speculation that Pluto will resemble Triton. The two are comparable in size, and Pluto is already known to have a thin atmosphere like Triton's, as well as a surface dominated by nitrogen ice. But there are also significant differences. For example, though smaller than our own moon, Pluto has *five* moons of its own, including a relatively large one (half the size of Pluto itself) called Charon, which has a surface dominated by water ice instead of nitrogen. While Pluto itself may end up showing some similarities with its possible cousin Triton, my bet is that the Pluto system overall will turn out to be just as new, strange, and alien as every other place that we've encountered in our travels out into the solar system. Planets, dwarf planets, moons . . . it really doesn't matter what we call them. They are a diverse, interesting, and just plain *cool* lot of neighbors that we share our solar system with.

With so many hopes pinned on the fast-approaching flyby of

Pluto, I can't help but think back to the *Voyager 2* Neptune flyby where almost instantly so much was revealed. And yet we all felt a mixture of exhilaration and regret as we looked in the rearview mirror, wanting more. Larry Soderblom recalls, "Our feelings as *Voyager 2* completed its last solar-system encounter and receded from Neptune toward the deep void of interstellar space were wistful and depressed. That evening I remember remarking to imaging team leader Brad Smith, on a positive note, 'Well, Brad, you only explore the solar system for the first time once.'"

The *Voyager 2* Neptune flyby would end up being the capstone to one of the greatest voyages of exploration ever conducted by our species. As I prepared to head back to Honolulu a few days after the Neptune flyby, I wandered one last time through the Science and Mission Operations areas of JPL Building 264. Someone had printed out a large-format image of one of *Voyager*'s final Neptune photos, a beautiful, parting view of the thin crescents of both Neptune and Triton as *Voyager* turned and looked back.

What was Charley Kohlhase thinking when he first saw that "last port of call" receding photo of the crescent Neptune and Triton, back in 1989? "I felt . . . what is it when you have an epiphany?" he says. "I felt nostalgia, I felt sadness that we were saying good-bye to the last worlds . . . but also great satisfaction. All the years that had gone by . . . and we had pulled it off. It was a big success. . . ." His words slowed, and he gazed upward, outward, focusing well beyond the confines of the room. "And to look at those limbs, of Neptune and Triton, as we're departing . . ."—tears welled up in Charley's eyes—"I won't ever forget that."

Jon Lomberg was similarly wistful about the end of one phase of *Voyager*'s mission and the beginning of the next. "One of the images

I've always imagined," he told me, "is riding along with *Voyager*, looking in the reflective gold surfaces of the record and seeing the volcanoes of Io, the cracks of Europa, the braided rings of Saturn, and so many other wondrous sights. It was bearing witness for all of us. It was bearing witness to all of these new worlds." Perhaps the simplest, most direct summary of *Voyager*'s influence on the many people touched by the adventure came from Rich Terrile: "*Voyager* was the most amazing experience of my life."

Part Three

LOOKING BACK,

LOOKING AHEAD

8

Five Billion People per Pixel

"Selfies" seem to be all the rage these days, and I am constantly amazed at the dexterity of my kids and their friends manipulating a smartphone held at arm's length to take these sometimes artfully framed self-portraits. I've also just recently learned about, and am fascinated by, the concept of people taking "dronies" of themselves and friends from remote-controlled cameras hovering above them on balloons, kites, or powered quad-copters. This fascination with self-portraits is not just a fad, however, and it extends far beyond just a subset of young, tech-savvy individuals. For as long as we've been looking upward and outward, we've also been looking inward, seeking clues that might help illuminate our place in the universe.

We tend not to think about the fact that the Earth is round, a sphere of rock and metal surrounded by a thin shell of air and water,

floating in space and moving swiftly under the gravitational influence of the sun. Moving swiftly? It doesn't feel like that at all, but the
numbers tell us otherwise: Earth is spinning at more than 1,000
miles per hour at the equator, and we're all traveling more than
67,000 miles per hour in orbit around the sun, and the whole solar
system is moving at about 450,000 miles per hour as the sun orbits
around the center of the Milky Way galaxy. We don't notice any of
these motions acutely, however, because gravity wins. The mass of
our planet and its gravity holds us, and the air and the oceans and
the mountains and everything else, to the surface, counteracting by
far any other forces that might be acting to fling us off. And *of course*
the Earth is a round sphere, right? We've all seen the pictures of our
Big Blue Marble suspended against the inky blackness of space that
the *Apollo* astronauts took on their round-trip voyages to the moon
back in the '60s and '70s.

But that knowledge was not at all obvious to most of the estimated 100 billion human beings who have lived before us, and who
may have imagined but never had the chance to see the results of the
simple experiment (simple once you've developed the technology
for space travel, that is): if you want to find out what the Earth is really like, look at it from the outside. The sixth-century BCE Greek
philosopher, mathematician, and astronomer Pythagoras of Samos
(the same guy of $a^2+b^2=c^2$ fame) is generally acknowledged as
among the first scientists to recognize that the Earth is spherical,
without, of course, the benefits of perspective provided by the modern space age. His evidence was indirect: Greek sailors saw southern constellations rising higher as they sailed south; when they got
really far south, the sun shone from the north instead of the south
(as it does north of the equator); and when the full moon passed

into the Earth's shadow during a rare lunar eclipse, the outline of the Earth's shadow appears curved.

It seemed obvious to Pythagoras. It would take more than 250 years, however, for another famous Greek mathematician and astronomer, Eratosthenes, to prove it and to accurately estimate our planet's size. He performed one of the most simple and famous scientific experiments of all time, and one that is easy for schoolkids to reproduce today, using just two sticks and a sunny day. One stick was in the southern Egyptian city of Syene (modern-day Aswan), on a day when, at noon, the sun was directly overhead and that stick did not cast a shadow. The other stick was in his own northern Egyptian city of Alexandria (Eratosthenes was the head of the Library of Alexandria, an amazing collection of all of the then-known books of the world—the equivalent of the Internet on Planet Earth in the third century BCE), where, on the same day, a stick would indeed cast a short shadow at noon. He knew that the angle between the sticks was the result of being at different places on a sphere, so he had an assistant (a graduate student, no doubt) walk off and measure the distance between Alexandria and Syene.

His predecessors Plato and Archimedes, not mathematical slouches, to be sure, used their best reasoning to estimate the diameter of the Earth as 14,000 and 11,000 miles, respectively. Eratosthenes, armed with data from his simple measurements, came up with around 9,000 miles, or within about 15 percent of the correct modern answer (7,918 miles). Not bad for sticks and shadows.

Fast-forward almost 2,200 years and we've entered an era when we can, in fact, just leave our planet, turn around, and take a look. The first time this was actually done was in the late 1940s, with cameras on suborbital German V-2 rockets that had been captured by the US Army after World War II and transported to the White Sands Missile Range in New Mexico. From altitudes of around 100 miles, the grainy V-2 photos showed the graceful curvature of part of the Earth's limb. However, the first truly "global-scale" photo of the Earth from space wasn't taken until nine years after the first Earth-orbiting satellites were launched. That photo, taken on August 23, 1966, by the NASA *Lunar Orbiter I* spacecraft, shows a beautiful, black-and-white crescent Earth appearing to rise behind the horizon of the moon.

Most people haven't heard of the *Lunar Orbiter* missions—a series of five robotic spacecraft sent to orbit the moon between 1966 and 1967 in order to scout for landing sites for the *Apollo* astronauts. They used a one-of-a-kind photo lab, born of the resourcefulness that has come to exemplify the space program. A set of Kodak film cameras was configured to take 70mm film pictures, to automatically develop the film inside a little chemical lab onboard the spacecraft, and then to digitize the developed negatives and transmit the digital data back to Earth. Essentially, the *Lunar Orbiters* used a scanner and a fax to send their pictures back, and they succeeded in mapping more than 99 percent of the moon that way. When the opportunity to take a photo of the Earth was recognized by *Lunar Orbiter I* mission controllers, they had to seek permission to take the risky step of using the spacecraft's onboard thrusters to tilt the camera's view toward the lunar horizon, where the Earth would be. It was risky because the Boeing spacecraft engineers pointed out that if it didn't tilt back to its original orientation, that would have effectively ended the mission, only shortly after it had begun. NASA did approve the maneuver, and the resulting photo is indeed spectacular. I think that officials at NASA headquarters and Langley

Research Center (which managed the mission) were compelled to approve the request, given the potential public relations value of what would be another first for America's space program, against the backdrop of important Soviet advances in lunar exploration and the looming deadline of landing astronauts on the moon "before this decade was out," established by a bold vision of the late President Kennedy.

The *Lunar Orbiter I* photo of "Earthrise" was indeed a huge public relations hit. It became an instant poster handed out by NASA to members of Congress and visiting dignitaries as an example of tangible progress toward the *Apollo* landing goal, as well as the prowess of NASA's young robotic exploration program in general. Just a few weeks later, *LIFE* magazine ran the photo in a two-page spread. Eight months later, NASA's *Surveyor 3* lunar lander one-upped the feat by taking the first color photo of the whole Earth from space—another beautiful crescent view. The public relations potential and raw visceral motivational power of viewing the Earth from space was apparent in these first early efforts.

The next big leap in viewing our planet from space came from the crew of the *Apollo 8* mission—the first full-up test of most of the components needed for a successful landing on the moon, and the farthest trip away from their home planet that any humans had ever taken. *Apollo 8* launched on December 21, 1968, and the crew and mission-support staff back in Houston successfully navigated the two-and-a-half-day cruise to the moon, and then fired the Command and Service Module's main engine to brake into lunar orbit. Commander Frank Borman, Command Module Pilot Jim Lovell, and Lunar Module (LM) Pilot William Anders then spent the next twenty hours orbiting the moon ten times, becoming the first humans ever to do so. Bill Anders was technically the LM pilot on the mission, but since *Apollo 8* didn't carry an actual LM for landing on the moon, and since he was a scientist by training, his job was mainly focused on acquiring photographs of the moon that could help in the study of its geology and in the analysis of potential landing sites for future *Apollo* missions. According to the NASA recordings and transcripts, on their fourth orbit around the moon, on December 24, 1968, Anders, Borman, and Lovell had the following exchange:

> ANDERS: Oh my God! Look at that picture over there! There's the Earth coming up. Wow, is that pretty.
> BORMAN: (joking) Hey, don't take that, it's not scheduled.
> ANDERS: (laughs) You got a color film, Jim? Hand me that roll of color quick, would you . . .
> LOVELL: Oh man, that's great!

The three men were the first people to observe an "Earthrise" from another world. Anders and Borman took a number of color and black-and-white photos over the next few minutes, but once the crew returned and the film was developed, the photo that Anders took first has turned out to garner the most press and public interest, partly because it is in color, and partly because it is so well composed, benefiting from the lucky timing of the event compared to their busy flight plan. The crew had also captured the world's attention just a few hours after the *Earthrise* photo was taken, with their Christmas Eve reading of part of the Bible's Genesis creation story to an enormous worldwide television audience.

Some commentators credit the *Apollo 8 Earthrise* photo with helping inspire the first Earth Day in 1970, and even with providing the impetus to propel much of the modern environmental movement into the mainstream. After all, standing on the surface of the Earth, it is easy to feel the vast and seemingly infinite nature of our natural world. But seeing the Earth as an isolated sphere floating in the emptiness of space really drives home just how limited our resources truly are. In *LIFE* magazine's 2003 compendium "100 Photographs That Changed the World," wilderness photographer Galen Rowell named *Earthrise*, which featured prominently on the cover of the magazine, as "the most influential environmental photograph ever taken." Despite perhaps being eclipsed in the media and pop culture a few years later by the now-iconic December 1972 *Blue Marble* photo of the full Earth taken by the crew of *Apollo 17* (arguably the widest-distributed photo in the history of photography), to me the astonishing nature of these first views of our planet from space has had a compounding effect. They demonstrate, graphically, the frailty and isolation of our planet, the vulnerability of our home world compared to the vastness of space, and the dawning of a new global consciousness—one that is still struggling to achieve critical mass—that recognizes the special responsibility that our species bears for the stewardship of this precious cocoon of life called Earth.

THE LITTLE SELFIE

The *Voyager 1* imaging team was responsible for what I think of as the next advance in planetary selfies, taking the first photograph of the Earth and moon together just a few weeks after launch. Mission and imaging team planners like the late Andy Collins of JPL realized that as the spacecraft was departing from Earth, once it got to a distance of over 7 million miles away or so, it would be possible to photograph both the Earth and the moon together in the same field of view. The imaging sequence was partly justified as an initial test of the cameras after the rigors of launch. Had the intense shocks and vibrations caused any damage or change in performance relative to prelaunch expectations? Did the various rocket stages and thruster firings generate any contamination that might be fogging up the lenses? Did someone leave the lens caps on? Indeed, imaging team instrument scientist Candy Hansen, who had just started working on the *Voyager* project in the summer of 1977, recalls that the sequence also had an important spacecraft systems function. *Voyager*'s scan platform was stowed for launch and had to be deployed to its correct position once the vehicle was out in space. However, after the deploy command was sent, the sensor that could confirm that the scan platform had deployed and latched properly into place failed, so there was no way to know for sure whether it had happened. Well, Candy recalls Andy Collins and others thinking, if they commanded an Earth-Moon photo *assuming* that the platform was properly deployed and the positions of the Earth and the moon in the photo were as expected, that should prove it. And indeed, it did.

I would work with Andy many years later on the development of the cameras for the Mars rovers *Spirit* and *Opportunity,* and Candy's story about that clever work-around is typical of the ingenuity and creativity that he brought to solving many other problems in robotic space exploration systems. Importantly, though, Andy and the rest of the *Voyager* imaging team knew back in the summer of

1977 that they had a chance to take the first photograph (in a long series of historic "firsts" bestowed upon us by *Voyager*) of the Earth and moon dancing together in space. Indeed, there was wide public interest in seeing, through the eyes of *Voyager,* our home planet and its nearest celestial neighbor from a completely new and different perspective than ever before. "A pretty pair," remarked Carl Sagan, showing off the historic *Voyager* photo in one of the episodes of his 1980 television series, *Cosmos.*

Three years later, as *Voyager 1* sailed past Saturn and was slingshot up and out of the ecliptic—the racetrack-like plane that the planets orbit within—its mission of photographic exploration was coming to an end. As planned, the spacecraft would keep traveling on an upward path, rising higher and higher above the rest of the planets. The cameras would be turned off, and the fields and particles experiments would take over as the primary science of the mission shifted to the exploration of the limits of the sun's influence on our solar system. The cameras were working fine; it was just that they used a large fraction of the slowly dwindling plutonium power supply's electricity, and besides, there just wasn't anything for *Voyager 1* to photograph after Saturn and Titan.

Carl Sagan thought differently. In his mind, perhaps in his dreams, there was at least *one* more historic "first" that the *Voyager 1* cameras could achieve, at least one more new perspective that we could—and he strongly felt *should*—embrace. *Voyager 1* was rising out of the ecliptic plane like an airplane rising off a runway, slowly revealing things that could not be seen so clearly, or at all, from the ground. Sagan and *Voyager* imaging team planners like Candy Hansen realized that it would be possible from *Voyager's* new perspective to take a portrait of not only the Earth, but also almost the entire family of the sun's planets. It would be the first solar-system selfie, a name that I think would have made Sagan smile.

But not everyone was a fan of the idea. For many years, spacecraft engineers and *Voyager* imaging team members had fastidiously avoided inadvertently pointing the sensitive Imaging Science Subsystem cameras at the sun. Their reasoning: the cameras used telescope optics to focus their images; sunlight accidentally piped down through those magnifying optics could heat up the photodetectors and fry the system. Pointing at the sun is *bad*. And so now Dr. Sagan wants to do what? *Point at the sun?* Who is this guy?

"But there's nothing else to look at," retorted proponents of Sagan's idea. "If we burn out the cameras in the effort, so what?" However, while *Voyager 1* had nothing else to take pictures of after the Saturn flyby in late 1980, her twin *Voyager 2* was steadily speeding on to a flyby of Uranus in 1986 and then, the team hoped, of Neptune in 1989. The camera systems on the spacecraft were identical, and so for calibration or diagnosis of certain potential kinds of problems, the cameras on *Voyager 1* could theoretically still be used to diagnose any software or hardware problems that might occur on *Voyager 2*. That backup functionality, even if unlikely to ever be needed, would not be available at all if *Voyager 1*'s cameras were fried in the attempt to take a solar-system family portrait.

Plus, the imaging team was dwindling in size due to post-Saturn budget cuts, and the people still working on the project were fully occupied preparing for the Uranus encounter. Sagan and others knew that they couldn't defend the solar-system portrait request on scientific grounds. "The point of such a picture would not be mainly scientific," wrote Sagan. "I knew that, even from Saturn, the Earth would appear too small for *Voyager*'s cameras to make out any detail. Our planet would be just a point of light, not even filling a single pixel, hardly distinguishable from the other points of light it could image from nearby planets and far-off suns. But I thought that—like the famous frame-filling Apollo photographs of the whole Earth—such a picture might be useful nevertheless as a perspective on our place in the Cosmos." What a perspective, indeed. But the need for the team to focus their efforts on *Voyager 2* meant that the "pretty picture" work would have to wait. And so Sagan and others waited, for a decade, while *Voyager 2* made spectacular discoveries at Uranus and Neptune and *Voyager 1* climbed steadily higher above the ecliptic. . . .

Finally, after the successful August 1989 flyby of Neptune and the completion of the playback of all the imaging data taken during approach, flyby, and departure, Sagan and others once again raised the issue of trying for the solar-system portrait. While *Voyager 2* had been diverted southward after gracefully arcing over the north pole of Neptune, it was still relatively close to the ecliptic. *Voyager 1*, now more than 30 degrees above the plane of the solar system and still climbing, had the superior view. In late 1989 the idea was pitched again, but once again it was put off while additional calibrations were performed on the cameras to make sure that the Neptune images, especially, could be properly analyzed. It seemed a prudent precaution, and so Sagan and the other advocates waited some more.

The possibility came to a head after the Neptune flyby, though, when it was revealed that because of impending budget cuts to the *Voyager* program, many of the technicians who were responsible for the commanding of the cameras and the pointing of the spacecraft were to be laid off or transferred to other jobs almost immediately.

These people's skills would be needed to plan, acquire, and process the portrait—which had to happen soon if it were going to happen at all. An internal debate began within NASA over whether the cash-strapped *Voyager* project could afford to spend time and effort on what some people apparently regarded as a superfluous stunt. In the minds of some of the leaders of the planetary exploration program at JPL and NASA headquarters at the time, not only was there no science value in these images, the attempt would be yet another distraction as the project was winding down its staffing and preparing for the long interstellar phase of the mission.

This kind of attitude was far from uncommon in the leadership of NASA at the time. So-called education and public outreach activities like Sagan's solar-system portrait were not regarded as worthy of inclusion on planetary-exploration mission budgets. The attitude was pervasive across much of science in the '70s and '80s and was part of the reason that Carl Sagan had been denied membership in the National Academy of Sciences. Much of his work was regarded by his peers, especially members of the Academy, as "soft" science—communications and education-related or even (as I can imagine in the minds of some of his more jealous peers) grandstanding.

Times have changed. NASA and National Science Foundation proposals for funding the research programs of individual scientists now *must* demonstrate how we will communicate and disseminate our results to the general public, and in what specific ways our work has general relevance and importance to our society.

Fortunately, after the Neptune encounter, top NASA officials such as Associate Administrator for Science Len Fisk and Administrator Richard Truly shared Carl Sagan's vision of the historic, aesthetic value of the solar-system family portrait. Ed Stone was also a strong supporter of the idea. He recalls a dinner at Caltech organized by Sagan and The Planetary Society just before the *Voyager* Neptune flyby in 1989, during which he, Sagan, Fisk, and *Voyager* Project Manager Norm Haynes talked about what it would take to make "the picture of the century" happen. By this point in time it was essentially a budgetary issue, as *Voyager*'s funding was set to ramp down steeply right after Neptune. Happily, Fisk and Truly interceded to make sure the people and resources were made available for this one last *Voyager* mosaic, which was taken on February 14, Valentine's Day, 1990.

Voyager imaging team liaison Candy Hansen was involved in the planning of the mosaic, and she recalls that even after the NASA HQ directive, there was lingering disdain from some of the project leadership. "Some of the science management was doing the equivalent of holding their noses," she recounted to me, "and so at our sequence kickoff meeting they made me—essentially acting as a sequence engineer—make the presentation, rather than the usual process of asking a science team lead. But that was my chance to take the podium and sermonize about what a fabulous opportunity this was and how profound the images would be—hah!" Ed's and Candy's support and Carl Sagan's enthusiasm made a difference, and with the blessing of NASA's top brass, they made it happen.

Even twenty-five years later the resulting mosaic is mesmerizing. Starting out at Neptune and working its way inward in case the cameras got damaged, *Voyager* was commanded to snap photos of one planet after another. When the camera was pointed toward the sun to try to photograph the inner planets Mercury through Mars—all bunched up close to the blinding glare of our parent star—there was some saturation of the images, but the camera was not fried. "In order to minimize the glare from the sun while trying to image the inner planets," Randii Wessen told me, "*Voyager 1* was commanded in such a way that its high-gain antenna blocked some of the glare for the cameras—like a person at the beach, placing their hand in front of their face to block the sun." It was Candy Hansen's job to look through the images as they came in, to make sure that everything had gone OK with the observation. Just like Rich Terrile, by this point in the mission, she knew in great detail what stars look like in *Voyager* images, and she also knew every little artifact and blemish of the *Voyager* cameras by heart. So even though the photos were mostly empty black space, she was still able to quickly zoom in and isolate the stars and the blemishes from . . . something else. "Found Neptune—check. Saturn—check. No Mars—as expected, the crescent view would be too dim," she recollects, remembering her first views of the solar-system portrait images. "I finally got to the photo that was pointed at the Earth. At first I couldn't find it—there was a lot of scattered light in the image—but then I spotted it, in a ray of that scattered light." It turned out that one swath of yellowish scattered sunlight had passed right through *Voyager*'s photo of the Earth.

Our world was "a pale blue dot . . . a mote of dust suspended in a sunbeam," Carl Sagan poetically intoned. "As I am sitting here recalling that experience," Candy wrote to me, "there are chills going down my spine, just like that day when I saw our little planet from a vantage point so far away."

Once again the citizens of Planet Earth, then numbering about 5 billion, bore witness to the next great paradigm-shifting change in perspective. This time it was not just off-world, not from just beyond our backyard, but a vista from *out there*, looking down from near the edge of the sun's realm to behold the entire solar system. As was his way, Sagan challenged and inspired us to internalize this new perspective, and to use it to guide our paths forward. "It has been said that astronomy is a humbling and character-building experience," he wrote. "There is perhaps no better demonstration of the folly of human conceits than this distant image of our tiny world. To me, it underscores our responsibility to deal more kindly with one another, and to preserve and cherish the pale blue dot, the only home we've ever known."

The powerful message and literal imagery of *Voyager*'s *Pale Blue Dot* photo led to a string of selfies by subsequent planetary exploration missions. Many of the people who have gone on to become leaders in NASA's robotic planetary exploration missions were either trained by, worked with, or became disciples of Carl Sagan, and as such they are the kind of people who understand the enormous symbolic, inspirational, and educational value of photographing our home world from space. Some of my favorite planetary selfies include hypnotizing movies of the Earth spinning gracefully on its axis while the *Galileo* and *MESSENGER* spacecraft made their gravity-assist flybys of our planet on their way to Jupiter and Mercury, respectively; a glorious HD movie of a full Earth rising over the horizon of the moon from the Japanese *Kaguya* lunar orbiter; and absolutely stunning photos of our distant, faint planet nestled against the rings of Saturn, taken while the *Cassini* orbiter was passing through Saturn's shadow.

For the most recent one of these spectacular *Cassini* photos, NASA and the *Cassini* imaging team asked people to go outside at a particular time of day (or night) on July 19, 2013, look up, and smile and wave as the *Cassini* camera, from nearly a billion miles away, took an image of the (now) 7 billion of us all crammed into a single pixel. *Cassini* imaging team leader Carolyn Porco, who worked with Sagan on the *Voyager* imaging team, wrote on her Facebook page, "After much work, the mosaic that marks that moment the inhabitants of Earth looked up and smiled at the sheer joy of being alive is finally here. In its combination of beauty and meaning, it is perhaps the most unusual image ever taken in the history of the space program."

There are many more spacecraft self-portraits of our home world—it seems like we can't stop wanting to look at ourselves from new perspectives. My favorite is one that I played a role in helping to take (I guess selfies are like that). I count myself among Carl Sagan's disciples, having been influenced at an early age by his *Cosmos* TV show, his books and magazine articles, and by the immense good fortune of getting to be a colleague of his for a short time at Cornell University. When I had the chance to help lead a robotic planetary-imaging investigation of my own as the lead scientist for the Pancam color stereo cameras on the Mars exploration rovers *Spirit* and *Opportunity*, I was looking for chances to take photographs that would capture some of the same aesthetic, artistic, and inspirational appeal as the *Pale Blue Dot*. Happily, my friend and rover team leader Steve Squyres, another Sagan disciple, was a kindred spirit.

One such chance came in March 2004. Both *Spirit* and *Opportunity* are solar-powered, meaning that the power available to drive or take pictures is often dictated by the amount of dust in the atmosphere or by whether the solar panels are dusty or clean. Rover team colleagues Mark Lemmon, Mike Wolff, and I were all originally trained as astronomers, and so we'd been looking for ways to take some astronomical photographs—that is, of stars or other celestial objects, with the rover cameras. About 63 Martian days or "sols" into *Spirit*'s mission in Gusev Crater, we found ourselves with an abundance of solar power, and thus able to power the rover and camera heaters that would let us take images in the extra-frigid twilight hours before sunrise or after sunset. We knew that Earth was a "morning star" as viewed from Mars at that time (just like Venus and Mercury are sometimes visible as "stars" during twilight from Earth), and we had the power—could we spot ourselves in the Martian sky? Twilight is bright because of all the high-altitude dust in Mars's atmosphere, so we weren't sure if we would be able to see the Earth against the five a.m. predawn sky. But the next day when the images were beamed back—voilà, there we were! We'd taken the first photo of our home world from the surface of another planet.

We one-upped our feat by taking the first Earthrise *movie* from the surface of another planet in late 2005, using the cameras on the *Opportunity* rover to snap the Earth and Jupiter rising gracefully in the predawn sky above the dunes of Meridiani Planum. We'd worked up valid scientific justifications for taking all these Earthrise photos—for example, the need to measure the thickness of dust or water ice clouds/fog in the early-morning Mars atmosphere. But in the end, for me, it was just the sheer thrill of being able to look up, to glance back—like the *Voyagers*—to take a historic and introspective photo from a rare perspective, and to ponder what it means to explore a world where *we* are the aliens, experiencing it vicariously through the eyes of a robot.

9

The Edge of Interstellar Space

WHERE DOES THE solar system end? When I was in grade school, there were nine planets, and once you got out to Pluto, that was it. For science fair one year I spray-painted the inside of a box black, speckled it with white paint for stars, tipped it on its side, taped a drawing of the sun on one end, poked nine holes in the top, and hung cutouts of the planets from pieces of yarn. Solar system in a box! Scientists back then didn't know a whole lot more.

The *Voyagers* and other robotic missions since have revealed the incredible diversity of worlds within our solar system, including what are essentially mini solar systems around Jupiter and Saturn. Some moons, like Ganymede and Titan, are larger than the planet Mercury. Those moons, plus Europa and Enceladus and others, may harbor subsurface oceans. Earth is not the most volcanically active place in the solar system—Io is. Some large asteroids, such as Vesta and Ceres, appear to have had geologic histories as active as any planet's. And—among the most surprising discoveries—the solar system does not end at Pluto. Starting in 1992, astronomers have been discovering more and more relatively large, Pluto-sized planetary bodies lurking in the Kuiper Belt beyond Neptune. Almost 1,300 KBOs are now known, and the census is far from complete. The largest one yet found is called Eris, and it's a planetary body almost 1,500 miles wide (larger than Pluto, and about one-quarter of the mass of the Earth) with a moon of its own, called Dysnomia. Eris orbits, on average, almost twice as far from the sun as Pluto—the solar system doubled in size again when Eris was discovered in 2005. Astronomers estimate that there could be 100,000 or more KBOs larger than a hundred miles or so across, and perhaps hundreds of millions of them that are more comet-sized, only a few miles across.

The glut of Pluto-sized bodies being recently discovered beyond Neptune is what gave Pluto itself all that trouble, of course. Rather than accepting the fact that there are indeed many hundreds of newly discovered planets out there, and countless more still to be found, some astronomers chose to be "splitters" instead of "lumpers." In 2006, after some contentious debate, the International Astronomical Union (IAU)—the world's governing body tasked with giving planets and moons and asteroids and comets (as well as craters and mountains on those worlds) their names—decided to strip Pluto, and other places like it, of their "planet" status. Instead, such worlds were demoted to "dwarf planet" status, and the number of true planets in the solar system was decreased to eight, throwing textbooks and elementary school science-fair projects into chaos and disarray.

I'm a card-carrying member of the IAU and generally proud and supportive of the work that my colleagues in that organization do on behalf of astronomy and planetary science worldwide. But this time, I think they got it wrong. Personally, I judge a planet (like a person) on what's on the inside, rather than what it looks like or where it's been. Mercury is a planet because it has had a complex geologic history, including formation of a core, mantle, and crust, and the eruption of volcanoes on its surface, all fueled by substantial internal heat. It happens to be in orbit around the sun. Io, comparable in size, has had a similarly complex surface and interior geologic history. It happens to be in orbit around Jupiter, but it is still *the same kind of object*. So I call Io a planet. As well as Europa, and Ganymede, and Callisto. Plus Titan, Triton, Enceladus, Dione, Rhea, Tethys, Ariel, Ceres, Vesta, Eris, our own moon, and lots more. And Pluto—for God's sake, it's got an atmosphere and *five moons of its own*. If that's not a planet, I don't know what is. By my reckoning (and I'm a bit of a weirdo among my astronomy friends for this), our solar system has about thirty-five known planets so far, and it's likely that dozens more will be discovered over the coming decades. Let's celebrate those numbers and the diversity of planetary characteristics within our cosmic neighborhood rather than splitting them up into categories implying substandard status, such as "moon" or "dwarf planet." I'm a lumper rather than a splitter.

SOLAR WIND

Although astronomers and planetary scientists don't yet know exactly how far toward the nearest stars the sun's *gravitational* influence extends (it's probably somewhere near a half to two-thirds of the way), they have been expecting over the past decade or so that far-flung spacecraft like *Voyager* should soon be able to find the edge of the sun's *nongravitational* influence on the solar system. The sun produces energy by the conversion of four hydrogen atoms into one helium atom deep in its interior, at super-high pressures and at temperatures of millions of degrees. The conversion releases a tiny bit of energy, in the form of photons and other subatomic particles like protons and electrons, that bounce around inside the sun and eventually make their way out. The sunlight—photons—that warms our faces on a sunny afternoon was created, on average, deep inside the sun, maybe 50,000 years ago or more. The stream of protons and electrons coming off the sun every second creates a flow of charged particles called the *solar wind*. The solar wind creates a giant spherical "bubble" around the sun in interstellar space, known as the *heliosphere*. The heliosphere extends far beyond the orbit of Neptune, until it becomes so diffuse and weak that it merges into the background of rarefied hydrogen and helium gas that permeates the space between the stars—the *interstellar medium*. The sun and every other star reside inside their own such cocoons, blowing bubbles in the interstellar medium from their own solar, or stellar, winds. Like all bubbles, there must be an edge, a boundary between inside and outside of that bubble. Inside the bubble is the solar wind, outside the bubble is the *interstellar wind*. Finding that edge, then, and going beyond it, provides a way to explore truly *interstellar* space.

In some grand and philosophical way, as *Voyager* plowed on toward the boundary of this bubble, it became important to be able to determine the precise moment in time when we could say without question that we had left the confines of our solar system and we were now "outside," in interstellar space, the space between the stars.

While the vast distances traversed by *Voyager* to date are nearly incomprehensible on any human scale, it is perhaps even more difficult to grasp the lonely future of our spacecraft as it travels through interstellar space, heading off into infinity. Yet despite the cosmic emptiness that we are facing, the dream of the Golden Record remains alive in our hearts and minds. We can imagine a time in this incomprehensible future when some vastly superior beings, traveling these distances with the ease of today's intercontinental airline flights, would receive an earnest message from an Earth long gone but preserved in small part aboard our timeless *Voyager* emissaries.

But before it could be declared that Voyager had crossed into interstellar space, particles and fields emanating from our sun as well as from outside of our solar bubble would have to be tracked so we could witness the changes directly. Solar astronomers have discovered that, like the winds of our own planet, solar wind streams are in constant motion, acting out their own solar weather systems. Sometimes the solar wind is gentle and flows smoothly, like a breezy day. The slow solar wind (where "slow" is only 900,000 miles per hour) is an extension of particles that were accelerated through the sun's upper atmosphere—the expanding "corona" of the sun. And the fast solar wind (at more than 1.7 million miles per hour) appears to stream off the sun's visible surface (known as the *photosphere*). A few billion pounds of material streams off the sun every second, but the mass lost over time has still been only a minuscule fraction of the overall mass of the sun. Although invisible to our eyes, we can see evidence for the solar wind in the beautiful ion tails of comets, which always point downstream in the solar wind, away from the sun. These somewhat steady breezes are interrupted by occasional gale-force storms of particles called *coronal mass ejections*—the giant, looping arcs of hot plasma gas that launch off the surface of the sun and send electromagnetic shock waves and sprays of ionizing radiation outward toward the planets. Sometimes these waves and radiation produce glorious auroral displays in Earth's polar regions, and sometimes they also wreak havoc with electronics in orbiting satellites and surface power grids. The sun has weather, and it's important for our modern, electronic civilization to pay attention to it.

That's where space physicists like Ed Stone come in. Ed has spent his career working to understand high-energy particles from the sun and other cosmic sources, how they interact with the magnetic fields of the sun and the planets, and what they can tell us about how the planets, the sun, and other stars work. Ed's Cosmic Ray Subsystem (CRS) instrument is designed specifically to measure the energies and intensities of high-energy particles from the sun and other sources in the galaxy, in order to map out the sun's magnetic field as a function of distance, and to understand the effects of that field on the planets. It's an example of "squiggly line science" (you know, like the medical devices they use to monitor your vital functions, or the seismometer plots you can see monitoring for earthquakes in some science museums). Ed's instrument generates streams of data that most often appear on plots and graphs, rather than images. But scientists like Ed read them with ease. If you want to study the geological diversity of a planet's surface, pictures are the way to go. But if your aim is to understand the energies and densities of subatomic particles in space, a picture simply won't do. Other sorts of measurements are necessary. And it's a little-known fact that some of the most important discoveries from missions like *Voyager*, the Mars rovers, and many other missions come not necessarily from the pictures but from the investigations that produce squiggly line science.

After the Neptune flyby, and the success of the *Pale Blue Dot* photograph, *Voyager*'s focus was shifted almost entirely toward the fields and particles investigations that helped to characterize the interactions of the solar wind with the magnetic fields of the giant planets. With our beloved planets fading into the distance, these became the only instruments with something left to measure. And not just *something*, but something profound. Where does the solar wind stop blowing? Where does the sun's influence give way to the different kinds of fields and particles that infuse the spaces between the stars? Finding this edge, the edge of the heliosphere's bubble known as the *heliopause*—and studying for the first time the nature of interstellar space—became *Voyager*'s prime directive.

Ed Stone's CRS instrument can tell the difference between high-energy particles (nuclei) that have come from the sun (solar energetic particles) and those that have come from outside the solar system—from elsewhere in the galaxy (cosmic rays). Some high-energy cosmic rays do make it across the heliopause boundary, and *Voyager* has been characterizing them for decades. However, many of the lower-energy protons and electrons that make up cosmic ray particles from elsewhere in the galaxy can't actually pierce the bubble of the sun's heliosphere, and instead they are diverted around the solar system, like water diverts around an island in a river. Indeed, trying to measure the full range of energetic particles that are characteristic of the interstellar wind, not just the solar wind, was a goal for Ed's investigation way back when the instrument was being designed and built in the early 1970s.

I asked Ed whether deep in the recesses of his wildest dreams he actually harbored any hope back then, back when he was dreaming up his cosmic ray experiment in the '60s or when he became project scientist in 1972, that at least one of the *Voyagers* would survive to make it to interstellar space, to measure that interstellar, galactic wind. "Well, we hoped. When we started this mission, we had, as one of the objectives, to get to 20 AU," he told me. ("AU" is the astronomical abbreviation for "astronomical unit," where 1 AU equals the average distance between the Earth and the sun, or about 93 million miles.) "No one knew where the boundary would be. We had to propose for a new mission extension once *Voyager 2* completed Saturn, and we called it the *Voyager* Uranus-Interstellar mission. It was one leg at a time. Then the next leg after Uranus was the '*Voyager* Neptune-Interstellar' mission. So 'Interstellar' has always been there, it's just that none of us knew how big the bubble really is! And none of us knew how long the spacecraft could last."

The "Interstellar" focus of *Voyager*'s postplanetary mission was not an accident. Ed believes that his own cosmic ray instrument, for example, was put on specifically for this possible, hoped-for extended phase of the mission, where interstellar particles could be measured in their native habitat. "Which is really kind of remarkable, when you think about it," he confesses. There is a modern penchant for cut-rate mission budgets and strict adherence to carefully crafted specific mission goals. Ed Stone adds, almost to himself, "I'm not sure that today that would happen."

Both *Voyagers* had already been accelerated to high-enough speeds, by launch and by their subsequent giant-planet gravity assists, to be on escape trajectories from the sun's gravity. *Voyager 1*, which was diverted northward during its flyby of Saturn in 1980, is traveling fastest, at about 10 miles per second (more than 38,000 miles per hour). *Voyager 2*, which arced over the north pole of Neptune in 1989 and was then diverted southward, is not too far behind, traveling at about 9.5 miles per second (more than 34,000 miles per hour). Three other spacecraft—*Pioneer 10* and *Pioneer 11* launched in 1972 and 1973, respectively, and *New Horizons*, launched in 2006—are also on escape trajectories out of the solar system, but they are all traveling slower than the *Voyagers*, which are leading the race to find the edge of interstellar space.

While we're no longer in contact with the *Pioneers*, *Pioneer 11* could leave the heliosphere within the next decade or so, but *Pioneer 10* will take much longer (perhaps thirty to fifty years or more) because it is traveling "downstream," along the extended "tail" of the sun's magnetic field. How far does that solar magnetotail extend? Ed says, "I'm not sure. Hundreds, hundreds of AU?" At first that may seem surprisingly long, but consider that the sun's magnetic bubble extends more than 100 AU in the "upstream" direction, where it's flowing against the current of the interstellar wind. Maybe it's not so surprising, then, that it could extend two or three times as far in the other direction, where it's flowing downstream in the same direction as the flow outside the bubble. "The edge of the sun's magnetotail is undoubtedly a ragged or filamentary kind of thing based on the shapes of the magnetotails of the giant planets, and it probably just eventually merges with the interstellar background," Ed adds. *New Horizons* is traveling upstream into the interstellar wind, like the *Voyagers*, but it, too, is probably twenty to thirty years away from leaving the heliosphere, based on its later start and slower speed compared to the *Voyagers*.

Ed Stone and other space physicists worked up computer

models and predictions for when these spacecraft might reach the heliopause. The models use information on changes in the strength and shape of the solar wind measured over many decades by the *Voyagers* and *Pioneers*. "During the '90s, we would have a meeting every few years," Ed says, "and there would be a histogram showing the range of predictions. And that histogram just kept moving out in time as *Pioneer* kept moving out in space, not yet crossing the boundary. It seemed like the edge of the bubble was always 20 AU beyond where we were! Things were changing slowly, but really, we didn't know. By the early 2000s, however, all the different ways of estimating the size of the heliosphere were converging on the first major boundary—called the *termination shock*—occurring at about 90, plus or minus 10, AU. So if we saw that in just a few years' more travel, we'd know that we were getting closer to exiting the bubble."

They also used images and other information from astronomers about the nature of similar "bubbles" seen around other nearby stars. Young stars embedded in so-called stellar nurseries—clouds of gas and dust such as the nearby Orion Nebula—are particularly useful because their darker stellar wind cocoons are easily visible against the brighter nebula gas and dust in the background. In places where the boundaries between the stellar and interstellar materials around other stars can be seen—for example, in Orion Nebula images from the Hubble Space Telescope—there appears to be a strong shock wave at the upstream sides of those boundaries (where those stellar winds are running into the background interstellar winds), almost like the shock wave near the nose of a supersonic airplane. Aeronautics experts, as well as astrophysicists, commonly refer to that leading shock wave as a bow shock (or just a bow wave, if the difference in fluid speeds is not as high), an analogy to the bow of a ship plying its way through the water.

Indeed, *Voyager* and other spacecraft measurements (going back to *Mariner 2* in 1964) have shown that the solar wind is supersonic, meaning that a shock wave should exist close to its leading, or upstream, edge. The direction of the sun's motion relative to nearby stars and to the overall motion of the Milky Way galaxy is well known, and so which direction is upstream is also well known. It just happens to be in the direction that both *Voyagers* are traveling. Ed and his colleagues predicted that a classic bow wave should exist at this upstream end of the heliosphere, since the relative velocity of the sun compared to the local interstellar medium was predicted to be about 15 miles per second. The fast-moving solar wind, plowing head-on into a fast interstellar wind moving in the opposite direction (like two rivers flowing into each other from opposing valleys), was expected to produce quite a strong wave front, perhaps even a shock wave (though they couldn't be sure). *Voyager 1* was heading closest toward the actual predicted position of the bow wave itself (along the "nose" of the wave front) and thus was predicted to be able to get there first. Exactly when *Voyager 1* might cross that bow wave was anyone's guess, however. Ten years after Neptune? Twenty? Thirty? No one knew where the edge of the heliosphere would be, but everyone knew that the *Voyagers'* plutonium power supplies wouldn't last forever. If the spacecraft hadn't crossed the line by the late 2010s or early 2020s, the power supplies might not last long enough to see it happen.

AN INTERSTELLAR MISSION

As both *Voyagers* sped on toward their interstellar destinies, the science and operations teams transitioned into a different kind of mission. The imaging team was essentially disbanded, now that there was nothing new left to photograph, and the cameras had been shut off. The same is true of the ultraviolet spectrometer team, although the instrument is still used to collect occasional "automated" astrophysical measurements of nearby interstellar hydrogen. Shutting off or curtailing instruments helps to save power, which is slowly dwindling on both spacecraft. While *Voyager*'s radioactive plutonium power supplies will take eighty-eight years to drop to half their power levels, more than forty years have now passed since that plutonium was produced, meaning that power is down to around 75 percent of maximum values. The most important thing this power is used for is to heat the computer, radio transmitter and receiver electronics, and the remaining instruments. If left to soak in the cold of deep space, the temperatures of those systems would quickly drop to just a few tens of degrees above absolute zero, causing solder joints to break, resistors to crack, or any of a number of other possible fatalities.

Shutting off instruments and scaling back mission operations also helps save money. NASA's entire annual budget allocation from Congress has averaged around $17 billion per year lately, or about 0.4 percent of the entire federal budget. Of that, all the science done within NASA costs about $5 billion per year, and of that, all the solar-system science—robotic planetary missions and data analysis, laboratory studies, technology development—has averaged about $1.5 billion per year. My colleague Casey Dreier at The Planetary Society has recently pointed out that that's about the same as what Americans paid for dog toys last year. Don't get me wrong—I love my dogs and I want them to have fun! But it's important to put the costs to the taxpayers of this kind of grand exploration into perspective. And of that $1.5 billion per year, it costs about *$5 million* a year to keep the *Voyager* missions going. A significant amount of money, to be sure. *Voyager* scientists led by Ed Stone, along with JPL Project Manager Suzy Dodd and her mission operations team work hard to justify that $5 million request every year. "*Voyager* has been through something like eleven major reviews of its extended, extended mission," says Suzy Dodd. Everyone involved takes it very seriously that they make sure to use these amazing far-flung laboratories as efficiently as possible, to push the frontier of human knowledge ever outward, and to learn as much as we can about the far reaches of our solar system. Spread out over the lifetime of the project, the total cost of *Voyager* has been about a dime per year for every American. That seems quite a bargain.

"Now that we're in interstellar space," Suzy Dodd says, "we've reached the rarefied air of being an untouchable spacecraft." The *Voyagers* belong to all of us, they represent all of us, they will speak to the ages for all of us.

After Neptune, mission operations became known at JPL and NASA HQ as the *Voyager* Interstellar Mission. The goal of the *Voyager* Interstellar Mission, according to NASA, is "to extend the NASA exploration of the solar system beyond the neighborhood of the outer planets to the outer limits of the sun's sphere of influence, and possibly beyond." An important part of that, both officially and in the dreams of scientists like Ed Stone, has been to search for and find the heliopause, the boundary beyond the outer limits of the sun's magnetic field and the outward flow of the solar wind, and to directly measure the interstellar fields, particles, and waves beyond the influence of the sun. In other words, to extend the reach of human senses into interstellar space. Quite a dream.

With no working cameras, the *Voyagers* may be blind, but they are nonetheless still capably feeling their way through the outer solar system. Five different science instruments are still being used, many almost daily since the start of the Interstellar Mission, to touch and smell and taste the distant heliosphere. These instruments measure the plasma ions in the solar wind ("plasma" is a physics term for an ionized gas consisting of positively charged ions and negatively charged electrons—a common example is the gas inside a fluorescent lamp); the compositions, directions, and energies of solar wind particles and interstellar cosmic rays; the strength and orientation of the solar or interstellar magnetic fields; and the strengths of natural radio waves that are thought to be originating from nearby interstellar space. One important part of one of *Voyager 1*'s instruments was not operating properly, however. The spacecraft's instrument that was designed to measure the density of ionized hydrogen plasma in interplanetary space had stopped working shortly after the Saturn flyby in 1981. There were other ways to indirectly measure the amount of hydrogen *Voyager 1* was encountering, but its inability to make a direct density measurement would lead to some controversy later.

Although *Voyager 1* had a head start after completing its planetary mission at Saturn in 1981 and was already forty times farther from the sun than at launch, the Interstellar Mission didn't officially begin until *Voyager 2* passed Neptune in 1989, at a distance of about 31 AU. Even at those ranges, *Voyager's* instruments were still measuring a steady stream of high-energy solar particles and magnetic fields heading out into space on somewhat "radial" trajectories—that is, radiating generally away from the sun, like air inside an expanding balloon. The fields and particles were not perfectly radial, however, but were instead deflected somewhat—as if responding to the looming edge of the heliosphere at some unknown distance ahead.

Since 1989, communications technicians at the NASA Deep Space Network facilities in California, Australia, and Spain have faithfully captured data from the spacecraft for six to eight hours almost every day, using the "smaller" 34-meter (111-foot-wide) DSN radio telescopes to capture the faint signals from the meager 23-watt transmitters on the *Voyagers*. By the time those radio signals travel for more than ten hours at the speed of light, across vast distances now more than 100 AU from Earth, that 23 watts has faded to only 0.0000000000000001 watts, or barely a flea's whisper. But the *Voyagers* do a good job of pointing their antennas right at the Earth, and the DSN does a good job of pointing its antennas right at the *Voyagers*, and the very narrow transmitter frequency is pretty far from Earthbound or celestial radio noise sources, and so—somewhat incredibly it seems—it all works. About every six months or so, the DSN points its even larger, more sensitive, 70-meter (230-foot-wide) radio telescopes at *Voyagers*, and the spacecraft are commanded to download a bunch of higher-quality twice-a-week plasma-wave-instrument "wide band" (higher resolution, more sensitive) data, which are used to infer plasma densities that have been stored on the 8-track tape recorder instead of transmitted in real time. Once the data are confirmed to have arrived safely at Earth, the tape recorders are rewound and the process starts again . . . year after lonely year. . . .

Voyager's fields and particles scientists predicted that the spacecraft should pass through several distinct parts of the heliosphere before finally popping out of the bubble and reaching interstellar space. The first new and different place they figured they would encounter would be a boundary known as the termination shock. Ed Stone likes to give public talks about *Voyager*, and he is an enthusiastic and engaging speaker. One of his favorite slides is his "heliosphere in my kitchen sink" movie, where he tries to describe, using a kitchen-sink analogy, the kinds of places that the *Voyagers* will visit on their way out. Here's how it works: Empty your sink, angle the faucet so that it points toward one side or the other of the drain (not in the middle, where I'm assuming your drain is), and turn the water on full-throttle. The water crashes into the bottom of the sink and fans out into a lovely circular disk of water maybe five or six inches across that is flowing *radially away* from the impact point. That's like the deep inside of the heliosphere's bubble, except in the real solar system, the solar wind and the sun's magnetic field are wound (by the sun's twenty-five-day rotation period) into a huge Archimedean spiral with arms that move outward roughly radially away from the sun. But then look carefully at the water on the side of that bubble opposite the drain. That water is starting to slow down, to thin out, and to change direction because of the slight upward slope of the sink. There's probably a turbulent little zone there where the water is bubbling and churning a bit. That place where the water stops being radial and changes speed and direction is the termination shock and it marks the transition to a new and turbulent part of the flow, where the water is turned around and heads toward the drain. In the actual heliosphere, the termination shock occurs where the solar wind changes speed and direction because of the pressure of the interstellar wind coming from outside the heliosphere. The region just beyond the termination shock is called the *heliosheath* (the skin or covering on the heliosphere), and no one knew how large that region would turn out to be, because no one knew how much pressure was coming from the outside. The next stop beyond the heliosheath, though, should be the actual edge of the heliosphere—the heliopause.

Over the first decade of their Interstellar Mission, both *Voyagers* measured the density of the solar wind slowly decreasing, as it was spreading out at greater and greater distances. In December 2004, twenty-four years after passing Saturn and twenty-seven years after launch, at a distance of 94 AU from the sun, *Voyager 1* noticed a sudden drop in the speed of the solar wind (from million-mile-per-hour supersonic speeds to "just" a quarter-million-mile-per-hour subsonic speeds) and a jump in the density of the solar heliospheric particles, like a traffic jam on a busy freeway. At the same time, the instruments were able to sense an increase in the strength of the sun's magnetic field. *Voyager 1* had crossed the termination shock. In August 2007, far to the south and at a distance of 84 AU from the sun, *Voyager 2* crossed the termination shock as well. Both spacecraft were now in the more turbulent heliosheath. Next stop: the heliopause. But when? A year? A decade? More? I figured there must have been a betting pool. But Ed Stone said, stoically, "No, no. There was no betting. Just the histograms that we kept track of with everyone's estimates. I don't even know if there's a record of who voted for what." Pity. I bet Ed would have won the kitty.

Despite the incredible speeds of both spacecraft, the missions appear to be going in relative slow motion because of the enormous distance scale of the outer solar system. The spacecraft are traveling *10 miles every second* (imagine *Voyager* passing through your neighborhood at that speed) but still took nearly a decade and a half beyond their last planetary encounters to pass the termination shock and to enter the turbulent heliosheath. During that long cruise outbound, Ed Stone and other *Voyager* scientists moved on to work on other projects, to other missions, or even to retirement. "I feel extremely fortunate to have become the project manager in 2010," recalls Suzy Dodd. "During the twenty-year period that I was off the Project, we were just sort of sailing out towards interstellar space, not really having a good idea of how far away that was. I owe my job to the previous project managers who constantly had to fight against the mission being canceled during that time. It wasn't easy to keep *Voyager* going this long, not just from a technical standpoint, but also from a financial standpoint. There were doubters out there in the early 2000s who wanted to cancel *Voyager*. Once we got through the termination shock, though, people thought—OK, we should be getting closer." Indeed, once they crossed that boundary, many people began paying closer attention to the mission again, watching for changes in the densities, energies, and directions of the fields and particles measured by the *Voyagers*, searching for the telltale clue that the next—the ultimate—boundary had been passed.

HELIOPAUSE

After crossing the termination shock and entering the heliosheath, *Voyager 1* continued speeding outward at more than 37,000 miles per hour for nearly seven more years. Slowly, over that time, the background intensity of "outside" cosmic rays (hydrogen, helium, and free electrons) slowly kept creeping upward. Ed and his team interpreted that as meaning that there was a higher intensity of cosmic rays outside the heliopause that they would eventually (soon?) encounter, but that some of these "outside" cosmic rays were still slowly leaking, or diffusing, into the heliosphere. When *Voyager* was deeper inside the heliosphere, these outside particles couldn't get that far in. But now, getting closer to the edge, they could start to sense more strongly the storm waiting for them on the other side. The magnetic field lines kept slowly turning as well, until by 2010 they weren't pointing radially outward from the sun at all—in fact, in places the field had been turned around completely and was now stagnant or even in places pointing back in *toward* the sun. *Voyager 1* had moved into a sort of magnetic doldrum. The environment was changing over the years, but so far only gradually.

Things started getting weird, though, rather abruptly, on July 28, 2012. On that day, Ed Stone's cosmic ray counter instrument on *Voyager 1*, at a distance of about 120 AU from the sun, measured a sudden and dramatic 50 percent drop in the kinds of solar energetic particles that had been seen for about a decade inside the heliosphere. At the same time, another counter measured a big increase in the cosmic ray particles formed outside the heliosphere, in the nearby galaxy. But then everything switched back a few days later, and the environment went back to more normal levels of "inside" and "outside" particles. What the heck was going on? Again, a few weeks later, in mid-August, the inside particles dropped off and the outside particles jumped up—but then again went back to normal a few days later. Things seemed to be bouncing around, and it was hard for the team to make sense of what they were seeing on their squiggly line plots. Ed Stone would later refer to this phase of the mission as the time when *Voyager 1* was "dipping in" and "dipping out" of the heliosphere, along a somewhat jagged edge. "I can still remember taking the data home every night, and putting the plots on the refrigerator," recalls Ed Stone. "I couldn't stop thinking about them, wondering what would happen next." Suzy Dodd remembers seeing Ed give a talk in summer 2012 where he showed that plot, telling people in the audience, "This is the first thing I look at every day when I get up in the morning. And you should do that too!" And then, on August 25, 2012, *Voyager 1* saw the inside particles typical of the heliosphere drop off steeply to zero—and stay there. The outside particles jumped up dramatically at the same time—and stayed there too. The solar energetic particles were gone, replaced by nearly 100 percent interstellar cosmic rays. Was that it? Had the spacecraft just suddenly fallen off the edge of a proverbial cliff on August 25 and tumbled out into interstellar space? "It felt like I was standing on the shore of a particle beach, and the water comes up," recalled Ed Stone. "You're standing there and a wave comes in and gets your feet wet, and then the water recedes, and then there's another wave that comes in, and it recedes, and then finally the next one comes in and that's it. The tide has changed, and your feet are in the water all the time."

I asked Ed if he and the team celebrated that event in some way—maybe popping some Champagne corks or throwing a party? Don't get me wrong, *no one* would ever refer to Ed Stone as a party animal, but if ever there were an occasion for a space plasma physicist, a straight-up squiggly line cosmic-ray kind of guy, to let his proverbial hair down and celebrate, surely this would be it? "Well, it was really quite remarkable," he said. "We were having a preplanned *Voyager* Science Steering Group meeting at JPL, timed to coincide with the thirty-fifth anniversary of the launch of *Voyager 1*. So we had a dinner scheduled, and much of the team out there, and the spacecraft obliged by crossing this historic boundary just the week before its big birthday party!" I pressed him about whether he had a personal celebration of some kind, though. "No, but maybe I should have. It's because, somehow, having waited for it for thirty-six years ... we just weren't sure. I really wanted some confirmation that we were out there." Sounds like he was having fun, though I never got a straight answer about the Champagne.

Ed Stone is a careful, skeptical guy, and he wasn't yet ready to declare victory. "We couldn't be *sure* yet, because we hadn't measured the plasma density, and we hadn't measured the magnetic fields yet, but from a 'particle' point of view, we felt as if we were at least connected to the outside somehow, even if we weren't actually outside." He knew that crossing the edge of the solar system was a big deal, and that they'd want to make sure that it had really happened. The funky dropouts in heliospheric particles during the month before August 25 were troubling—what were they caused by? Was the edge of the heliosphere moving in and out (like the water on a beach) rather than sharp? Or had they entered some unknown, unexpected, strangely depleted region of the heliosphere that was still upstream of the edge itself? That would be an exciting discovery too. No one knew.

Because the cosmic-ray measurements could be interpreted in several ways, Ed and colleagues went looking for more clues to where they really were in other *Voyager 1* data sets. Most of their conceptual cartoons and computer models predicted that when the spacecraft crossed the heliopause, there would also be a sudden change in the direction of the magnetic fields—from the bunched-up, stagnant, turned-around fields measured just inside the heliosphere boundary to the more freely streaming fields of interstellar space in this particular part of the galaxy. But disappointingly, the direction of the magnetic field didn't change at all on August 25. Curious—the magnetic-field data said that they *hadn't* crossed out of the heliosphere after all. The clincher measurement would have been the density of the ionized plasma, because everyone agreed that the density should jump somewhere between 50 to 100 times higher once *Voyager 1* passed into interstellar space. But the plasma density instrument had broken back at Saturn in 1981, so there was no way to make that direct measurement. They couldn't be *sure* that they crossed the threshold. Cautious Ed Stone couldn't be absolutely sure.

Ed and the other *Voyager* fields and particles scientists now had a major conundrum on their hands. They couldn't prove that *Voyager 1* had crossed into interstellar space, but they had definitely crossed into *some kind* of new region, different from any that it had ever traveled in before. They struggled with what to do, what to report to their colleagues in space physics and to the rest of the world that was expecting some exciting news from *Voyager*. In the fall of 2012, armed with the information that they had in hand and mindful of the pressure from NASA, the public, and the media to report on what *Voyager 1* was experiencing, they decided to take the middle road. *Voyager 1* had certainly entered a region with a dearth of the solar heliospheric particles that had been seen before, so they could report that they had, indeed, passed into some sort of *depleted* region of space. And the lack of any significant change in the magnetic fields between the "normal" heliosphere and this new, depleted region suggested to some researchers that there was a connection across this transition zone. Ed and colleagues coined the phrase "magnetic highway" to describe the idea of relatively seamless, high-speed magnetic-field connections across this new, mysterious boundary. Pulling all the available data together, Ed and the *Voyager* team went to press with a series of peer-reviewed research papers that appeared in *Science* magazine in June 2013, hypothesizing the discovery by *Voyager 1* of a new region of perhaps interplanetary, perhaps interstellar, space—but a previously unexplored region, regardless. From a purely *particle* perspective, the spacecraft appeared to be outside the heliosphere. But there was no proof in the plasma, at least not yet. The *Voyager* team consensus in 2012— driven strongly by Ed Stone's need for definitive proof—was that they had not yet necessarily crossed the boundary. Maybe, but maybe not.

Outside of the *Voyager* team, there was both concurrence and controversy. Certainly most researchers in the field wanted to see the smoking-gun evidence that could have been provided by the plasma density measurements (if the instrument was working), and agreed that a conservative, wait-and-see interpretation was warranted. But others had already been convinced by *Voyager 1*'s data that the spacecraft had crossed into interstellar space. For example, a group from the University of Maryland and Boston University led by space physicist Marc Swisdak used the *Voyager 1* magnetic field measurements to develop a new computer model of the heliosphere that envisioned the heliopause as a "porous, multi-layered structure threaded by magnetic fields." In their computer simulation of *Voyager*'s flight, published in August 2013, the spacecraft had indeed crossed the heliopause and was now in interstellar space. "We think we are outside the heliopause," says Swisdak in an interview for *Science* magazine. But, he adds, in order to explain the big difference in cosmic rays but the unchanged magnetic field direction, "the boundary is very different than we thought." He concludes, "The very nature of the heliopause may come into question."

Meanwhile, in the months since *Voyager 1* had passed *whatever* important boundary it passed in August 2012, Professor Don Gurnett of the University of Iowa, leader of the *Voyager* Plasma Wave Subsystem (PWS) investigation team, knew that there was an indirect way to measure the density of the plasma in this new region of space but that the team would have to get lucky to measure it. Gurnett's instrument measures the size of waves that travel through the ionized atoms and molecules in the magnetic fields of the giant planets and in the solar wind, providing information on the density and temperature of those regions of space. During the Jupiter and Saturn flybys, *Voyager 1*'s PWS instrument could characterize the space environment well because of the waves of energy created by those planets' powerful, rapidly rotating magnetic fields. But while quietly cruising through the outer heliosphere, there were no such powerful disturbances to create waves in the ionized gas. At least, not often. Every once in a while, though, Gurnett and others knew, an enormous burst of energy from the sun, from a solar flare or so-called coronal mass ejection event, would spew forth out into the solar system, moving outward at high speed and making waves in the plasma. So, if the sun cooperated, perhaps they would see a giant flare make some waves in the *Voyager 1* PWS data, and the nature of those waves would tell them whether they were in an environment of low (solar system) plasma density or a high one (interstellar space). They would have to be patient and lucky to observe such an event from the sun. But since *Voyager 1*'s main plasma measurement instrument was broken, they had little choice but to wait and hope.

Gurnett's team did see a weak solar flare event pass by *Voyager 1* in real-time data radioed back in October-November 2012, but its effects on the plasma were too small to yield a good answer on the density. But then, in April-May 2013, the sun provided a remarkable and unanticipated gift to the *Voyager 1* team: particles from a very large and energetic flare passed by the spacecraft and created strong, easily measurable waves in the surrounding ionized gas. Later, the earlier fall 2012 event was also detected in the higher-sensitivity recorded data that were part of the regular (every six months) transmission of data back from *Voyager*'s tape recorder. The electrons were moving back and forth along the magnetic field—like sound waves compressing and uncompressing in an atmosphere—resonating with a frequency that told Don Gurnett and colleagues that *Voyager 1* was in a region of space with 80 times the density of ionized particles as in the solar system's normal heliosphere. Since the very *definition* of the heliopause—the edge of the heliosphere—is based on such a jump in density, it was a eureka moment. "When we saw that, it took us ten seconds to say that we had gone through the heliopause," he remarked.

Finally, Ed Stone and the rest of the *Voyager* team had the proof they needed, thanks to the far-flung effects of a rare, giant solar flare. There was no longer any need for waffling or conservatism, or for mysterious new depletion zones or magnetic highways—it was now official: *Voyager 1* had left the solar system. Don Gurnett and his colleagues published their results in a paper in *Science* in September 2013 that proclaimed the historic achievement to the world. "Now that we have new, key data, we believe this is mankind's historic leap into interstellar space," said Ed Stone at a press conference called to announce the discovery. "The *Voyager* team needed time to analyze those observations and make sense of them. But we can now answer the question we've all been asking: 'Are we there yet?' Yes, we are." Humanity's first baby steps beyond the influence of our own star had been taken—and we still had a capable, functional spacecraft out there (and another not far behind) to study interstellar space for the first time.

At least, that's the happy-ending Hollywood version of the story. There is, in fact, still some controversy and uncertainty—primarily from outside of the *Voyager* team but some even from inside—about whether the spacecraft has truly left the heliosphere. "I don't think it's a certainty *Voyager* is outside now," wrote space physicist David McComas of the Southwest Research Institute in September 2013. He and other colleagues remain puzzled by some inconsistencies in the available data. "It may well have crossed," he concluded, "but without a magnetic field direction change, I don't know what to make of it."

Indeed, George Gloeckler, a space physicist from the University of Michigan and an original member of the *Voyager* team since the start of the project in the 1960s, has stated flatly, "We have not crossed the heliopause." He and Michigan colleague Lennard Fisk have developed a model of what *Voyager 1* has measured that they claim can be explained by the continued piling up and compression of particles and magnetic fields within the heliosphere, behind a yet-to-be-crossed heliopause that still lies ahead. Gloeckler confessed in a *Science* interview, "We're way out there, by far a minority, but we can explain every *Voyager* result in a pretty natural way," based on their solar wind pile-up model. "That's quite a different story than typical models of the heliosphere," countered Ed Stone when I asked him for a reaction to the ongoing skepticism about whether *Voyager 1* has indeed left the heliosphere. But he seems open-minded. "If they're right, that will change our understanding of the kind of physics that is involved in these interactions between stars and their surroundings."

"We're learning what's out there," he continued diplomatically, reacting to another set of competing hypotheses that invoke a rattier, more turbulent heliopause boundary consisting of tendrils of extended heliosphere extending into the interstellar medium. "If we're really inside some strange extension of the heliosphere, what some colleagues might call a 'flux tube,' then it's a really big one because we've now been in it for more than two years," Ed explained. "So in some sense, that model of the edge of the heliosphere is different than the model that most people envision." Yet he remains gracious and diplomatic to a fault. "But if they're right, then maybe such flux tubes are important. Maybe they're a typical feature of the interaction of a stellar magnetic field and the interstellar magnetic field. Rather than a simple cometlike bubble, maybe there are these strange regions where the magnetic fields are connected." It's that kind of collegial open-mindedness that has made Ed Stone the natural scientific father of the *Voyagers* for more than forty years.

I had a chance to meet Jamie Sue Rankin recently, a second-year Caltech graduate student who is working with Ed Stone on the analysis and computer modeling of some of the *Voyager* cosmic ray data. Jamie is Ed's only current grad student, and she counts herself blessed to be working with *Voyager* data during such an exciting time in the mission, as compared to the last couple of decades when there really hadn't been much going on during the outbound cruise. "I moved to Pasadena in September 2012, roughly a week after *this* happened," she says, pointing to the huge drop in "inside" heliosphere particles that Ed had been tracking on his refrigerator plot. What timing!

Jamie was born in 1988, eleven years after the *Voyagers* launched and just a year before the Neptune encounter, and I asked her how it feels to be working on a space mission that is *much* older than she is. "That is a strange thing . . ." she says. "It has a technology that I haven't even seen before. I mean—magnetic tape recorders? I've never even pushed Play on a magnetic tape recorder!" She made *me* feel old. During that semester that she arrived at Caltech, she followed Ed through the whirlwind of team meetings and debates about whether they had crossed the heliopause. "When I walked into that first team meeting, I was definitely the youngest person in the room by at least twenty years," she recalled. But she said it was a great environment, and that they were incredibly supportive and eager to welcome a new person into the field. One of the managers even suggested that they should put Jamie into one of the press briefings with a mohawk, to try to help get younger people interested in the mission (and following in the footsteps of the *Curiosity* rover's famous "Mohawk Guy," JPL systems engineer Bobak Ferdowsi).

A mohawk wouldn't seem to be Jamie's style, though. She's a serious young researcher focused on using *Voyager* and other measurements to make PhD-quality discoveries about the sun's interactions with interstellar space. She works closely with Ed Stone and seemed genuinely amazed at the amount of time that he devotes to mentoring. "He is very, very patient," she says when I ask her to characterize her famous dissertation advisor. "He doesn't micromanage.

He doesn't get surprised by much, because he's seen *a lot*. He's got to be one of the busiest people I've ever met, but when we meet to talk science, he's never rushed. There is a trust there between us." It was the kind of sentiment about Ed that I had heard from others on the *Voyager* science team as well.

Jamie's enthusiasm for the future of the mission is exciting to soak in. "*Voyager* is like the Energizer Bunny—it keeps on running!" she says, with a touch of amazement in her voice. Aware that the team is aging and retiring, she takes her responsibility as a sort of "heir" to Ed Stone's part of the continuing *Voyager* empire seriously. "Somebody's going to have to run it in the future. Somebody's going to have to learn how to operate it from the experts who are running it right now. And somebody just has to have *faith* that it has come this far for a reason. I'm sure some people thought that *Voyager* would never last this long, that it would never get to interstellar space. 'Let's just turn it off,' they probably thought. It would have been so easy! But now we're getting these interesting results because of that faith, and that's why I'm here."

Like Ed, Jamie believes that *Voyager 1* has left the heliosphere, telling me that "it would be quite strange to imagine some sort of connected region where all these interstellar particles somehow get in, while the magnetic fields haven't changed. I don't think there's a real debate. I think part of the problem is that a lot of these competing models just can't resolve the details of the heliosphere on the same scale that *Voyager* can see them." Ed had told me earlier that people are starting to work on explaining smaller structures, like the ones *Voyager* is observing, but it's still an active, ongoing area of research. Jamie gets the chance to meet with many members of the space physics community who are thinking about these problems, either at *Voyager* team meetings or when they come to visit Ed at Caltech, and she told me that she asks them what they think. Almost without exception, she says, they tell her, "If Ed has said that it's left, then it's left." She went on to praise his careful, methodical style. "Ed is very wise about how to approach these kinds of discoveries. He doesn't jump the gun. The fact that he's gone through the process, being skeptical, and then changing his mind because of the solar flare results—there's a reason he's changed his mind. He's been analyzing data from space missions for fifty years, so I think his judgment on this is probably pretty much the best out there."

The uncertainty over the interpretation of *Voyager 1*'s data may never be fully resolved, partly because the spacecraft has only a partially functional set of instruments. But luckily, there is another similar spacecraft, but this one with a fully functional plasma-density instrument, just a few years and a few dozen AUs behind *Voyager 1*, that could resolve any lingering controversy and in the process become the *second* human-made object to leave the solar system. *Voyager 2*, which is heading outward on a much more southerly track than *Voyager 1*, passed through the termination shock in late 2007 and is now exploring a different part of the heliosheath, but still searching for its edge.

"We measure the plasma directly with *Voyager 2*, and so soon we'll *know* what kind of a discontinuity there is between the inside and the outside," says Ed Stone. "In fact, we already know that the plasma flow inside the heliosphere at *Voyager 2*'s location along the flanks of the heliosphere is totally different than it was along *Voyager 1*'s path along the nose." Specifically, the solar wind stagnated as *Voyager 1* approached the boundary, partly due to the lower solar wind pressure as the sun was going through a minimum in its cycle of activity.

"On *Voyager 2* we haven't seen any such slowing of the solar wind. We see it turning, as it has to, as it starts feeling the effects of the impending 'wall' of the interstellar wind," explained Ed. Will *Voyager 2*'s plasma density instrument eventually reveal the large predicted jump at the putative heliopause boundary? Will the sun's magnetic field lines smoothly merge into the interstellar field, like *Voyager 1*'s results and some new models of the heliosphere imply, or will there be a sharp change in those field directions, as had been predicted earlier based on classical models of the heliosphere? "I don't think anybody should take it for granted," counseled Ed Stone, "that Nature can't throw us another curveball. I'll be surprised if there aren't surprises!"

In the meantime, Ed has started putting his cosmic-ray intensity plot for *Voyager 2* back on his refrigerator. He had a copy pinned to his office wall when we met recently. For now, he explained, pointing to the squiggly line, "*Voyager 2* is just up and down, small variations, no big jumps yet like we saw with *Voyager 1* in the summer of 2012." I asked him when he thought it would cross out of the bubble. "It could be anytime. Probably a few more years. Who knows. We're watching. We're waiting." Suzy Dodd told me that she "feels like the spacecraft has given me several once-in-a-lifetime events, first with the Uranus and Neptune encounters, and now their crossing of the heliopause."

10

Other Stars, Other Planets, Other Life

FOR ALMOST FORTY years our metal and silicon emissaries named *Voyager* have been speeding away from the people who launched them, first flying past the giant planets and their gaggle of icy and rocky moons, and since then sailing out into the hinterlands beyond the sun's influence, where the once-familiar solar wind gives way to a different, unexplored interstellar wind. Once *Voyager 2* also passes the heliopause—the boundary between the solar and interstellar winds—both spacecraft will be functioning, but they won't keep working forever. The plutonium nuclear power supplies on the *Voyagers* generate electricity for the spacecraft's heaters, computers, and instruments at a very predictable power level. Over time, especially over the decades, that power level has slowly been dropping (from a total power level near 470 watts at launch to around 250 watts now) as the radioactive plutonium-238 slowly decays to nonradioactive lead-206.

The pioneers in our understanding of this magical alchemy called radioactivity were the physicists Marie and Pierre Curie and their colleague Henri Becquerel in France who, when studying certain kinds of phosphorescent minerals around 1896, found out that uranium-bearing salts spontaneously emitted their own radiation. They had discovered radioactivity, and the genie was out of the bottle. The reason that radioactivity isn't like alchemy ("turning lead into gold"), though, is that only certain starting atoms, such as uranium, have the correct unstable collection of protons, neutrons, and electrons that spontaneously (that is, with no human intervention or added energy of any kind) lose energy and turn into different, and eventually stable, elements. The predictable and steady—and for many elements, very slow—rate of decay of radioactive "parent" elements to stable "daughter" elements is what makes radioactivity such a great natural clock. Indeed, the decay of some radioactive elements takes billions of years, making these clocks inside rocks excellent natural ways to estimate the ages of samples of the Earth, the moon, and meteorites studied in the laboratory. In an interesting parallel, this phenomenon was elegantly applied when a small spot of radioactive uranium-238 was electroplated onto the cover of the *Voyager* Golden Records as a sort of timepiece, indicating to any extraterrestrial recipient who could measure the amounts of parent U-238 and its daughter radioactive-decay products the precise time that had elapsed since the spacecraft was sent out on its journey.

The element plutonium (atomic number 94, with 94 protons and 114 neutrons) has about twenty known radioactive forms, or isotopes. It and the next-lightest element, neptunium, had been discovered in 1940 as by-products of a uranium nuclear reactor. As the next two elements were discovered beyond uranium on the periodic table, physicists decided to name them after the next planets after Uranus: Neptune and Pluto (then, and to some still, a bona fide planet). The first isotope of plutonium to be intentionally manufactured in the laboratory, plutonium-238, was created in a nuclear reactor in 1941 by UC Berkeley physicist Glenn T. Seaborg and colleagues. They recognized Pu-238 as special because it generates a lot of heat when it decays radioactively but does not generate as many harmful gamma rays and other high-energy particles as other radioactive elements. This makes Pu-238 safer and easier to work with and especially useful in RTGs like *Voyager*'s.

In high school we learned about *Voyager*'s plutonium power supplies and the role that physicists had played in developing ways to power spacecraft far beyond the distances where solar panels would work. Plutonium is element 94, one of the "trans-Uranian" elements (heavier than uranium), radioactive, and one of a slew of very heavy elements on the bottom of the periodic table, only relatively recently discovered. In 1980 my best friend, Bob Thompson, asked our chemistry teacher how many elements were then known, and Dr. Manley replied that the best way to get the inside scoop would be to write a letter (remember, this is in the days before e-mail) to Dr. Glenn T. Seaborg at Berkeley and ask him about it. So he did. We all thought it was hilarious because Bobby's "reward" for asking a question in class was more homework. A few weeks later, though, to our and Dr. Manley's amazement, Seaborg wrote back to him. "One-hundred-six elements are now known of which the last (element number 106) has not yet been given a name," wrote Dr. Seaborg. It was a fairly short reply, but just the fact that a Nobel

Prize–winning physicist who had discovered *ten* elements on the periodic table took the time to write a personal letter was enough to get Bobby's picture in the local paper—VALLEY STUDENT HAS NOBEL PEN PAL read the caption. Bob has gone on to a career in astronomy studying giant stars and designing instruments for airborne and ground-based telescopes. In 1997, two years before Seaborg died, they named "element 106" Seaborgium, in honor of Bobby's Nobel pen pal.

DELIVERY TIME

Even though the plutonium on the *Voyagers* has decayed by only about 25 percent of its starting amount, the spacecraft are already starting to feel the pinch of looming power limitations. Even if there were something useful to photograph, for example, the cameras can no longer be turned on because they would gobble up too much power to also be able to run their heaters. The five remaining instruments use less power, but the power needs of the radio transmitters and the heaters are relentless. In addition, the spacecraft need to use tiny amounts of thruster fuel to accurately point their antennas at the Earth, and that thruster fuel is a consumable, and dwindling, resource. However, amazingly, Suzy Dodd says that only just recently did they finally switch to *Voyager*'s backup thrusters, because the primary thrusters were getting close to their expected lifetime limits, "after nearly 350,000 thruster cycles and thirty-four years of flight!" she said. The backups, which had never been used, are working just fine.

Team members predict that the spacecraft will have enough

power and thruster fuel to stay in communication with the Earth and operate at least one instrument until sometime around 2025, when *Voyager 1* could be more than 160 AU from the sun (more than 15 *billion* miles away), and *Voyager 2* could be out beyond 135 AU. By cycling off some of the remaining instruments and systems—those needing the highest power—when not in use after 2020 (or, at some point, off forever), mission controllers may be able to push the spacecraft's lifetimes beyond the mid-2020s. But eventually, the power levels will drop to critically low values where some of the heaters and other engineering subsystems will have to be shut off, and then the science instruments will fail or have to be shut off, one by one, with the lowest-power instruments like the magnetometer likely staying on the longest. Even then, though, according to Suzy Dodd, it might be possible to continue to operate the *Voyagers* "with just an engineering signal. We've been talking with the DSN about that possibility." That is, it might be possible to just stay in occasional radio contact with them well into the 2030s.

> *Earth to Voyager . . . still there?*
> (long pause)
> *Still here . . .*
> *Very well. Carry on. Talk to you soon.*

There is some possible science that could come from simply monitoring the strength of that faint signal, beamed back over such vast distances. According to Ed Stone, "As long as we have a few watts left, we'll try to measure something." Randii Wessen says that no one really knows how long the spacecraft will keep going. "I started at JPL in 1980—at Saturn—as an intern of the *Voyager* science support team. I always thought that the mission would end sometime during my professional career. Now I'm not so sure." Suzy Dodd has a specific goal: "We launched in 1977, and so if we can keep in contact, still doing science, until 2027, that would be fifty years. That's my goal—to have *Voyager* operate for fifty years." She and the *Voyager* team continue to have to fight to justify new NASA funding every few years. At some point, though, even the so-called engineering signal will cease from *Voyagers*, and they will embark on their final mission, to carry forth the Golden Record for all of humanity.

I hope that we stay in touch with them in engineering signal mode for a long time after science measurements end. Even just the simple act of pinging them by radio and waiting the hours—and eventually days—that it will take for the ping to be acknowledged and sent back, can teach us something about where they are and what it's like there. An interesting example of that comes from the precursor deep-space missions to *Voyager,* the *Pioneers.* After their encounters with Jupiter and Saturn, both *Pioneer 10* and *Pioneer 11* embarked on their own interstellar missions, heading out of the solar system in different directions from the *Voyagers: Pioneer 10* heading "downwind" in the heliosphere and close to the plane of the planets, and *Pioneer 11* heading "upwind," like the *Voyagers,* but only slightly above the plane of the planets. NASA's Deep Space Network kept track of the *Pioneers* long after their planetary flybys, until their plutonium-based power supply systems ran out of enough juice to power the radio and other critical systems in 1995 (when contact was lost with *Pioneer 11*) and 2003 (*Pioneer 10*). Subsequent analysis of the *Pioneer* radio signals revealed something curious, however. The spacecraft were not as far away from us as they should have been—something was slowing them down, by a tiny amount, year by year. It certainly wasn't from the gravity of any known objects, as that was being properly accounted for, or from any other obvious known forces. Perhaps it was some kind of new physics that could only be discovered by a long, lonely trip through deep, nearly empty space? No one knew, and the discrepancy became known as the *Pioneer* Anomaly.

Over more than twenty years, astronomers, physicists, and spacecraft engineers tossed around hypotheses about the gravity of small bodies like KBOs, or dark matter, or some other cosmological effect, causing the *Pioneers'* deceleration. Or maybe drag from particles in the heliosphere, or small helium gas leaks on the spacecraft that acted like mini thrusters, or some other spacecraft-related effect that hadn't been properly accounted for. Eventually after much head scratching, physicists and spacecraft engineers finally solved the *Pioneer* Anomaly. With funding from The Planetary Society, they tirelessly sifted through nearly thirty years of *Pioneer* tracking data, some of it recovered from ancient magnetic tapes restored to modern digital data files with funding from Planetary Society members. They solved the mystery. The deceleration turned out to be from a tiny, almost insignificant force created by heat ("thermal photons") leaking out of the plutonium power generation unit in a specific direction that happens to be opposite the sun based on the design of the spacecraft components. This tiny force directed away from the sun causes the spacecraft to recoil (Newton's "equal and opposite reaction") toward the sun, ever so slightly, slowing it down by the tiny amount observed. Pretty old physics, actually, but it took modern-day spacecraft forensics work to track it down.

Eventually, as the *Voyagers* continue on in pursuit of the limitless expanse of interstellar space, they will leave not only the realm of the solar wind but the realm of the sun's gravity as well. Both are traveling faster than the escape velocity of the solar system, presently some 128 and 105 AU from the sun, respectively. The sun's gravitational influence is predicted to extend one-third to halfway to the nearest stars, or maybe around 100,000 AU (or about 1.6 *light-years* away).

Out there is a hypothesized spherical swarm of comets and asteroids that have been cast out of the inner solar system by encounters with the planets or the sun over their lifetimes. Every few years a new comet on a long, elliptical trajectory is discovered; some of them, like 1995's Comet Hale-Bopp or 1996's Comet Hyakutake, produce spectacular displays of gas and dust as their ices are boiled off by the sun's heat into beautiful, gracefully arcing tails. Tracing the orbits of these and similar so-called long-period comets back to the outer solar system tells us that they come from enormous distances, and from any possible direction in the sky. This is what led Estonian astrophysicist Ernst Öpik and Dutch astronomer Jan Oort to hypothesize that the solar system is surrounded by a vast spherical shell of perhaps a trillion or more asteroids and comets, which we now call the Öpik-Oort Cloud (or usually, just the Oort Cloud), extending out to the edge of the sun's gravitational influence.

At their current speeds—and at 10 miles per second, they are the fastest objects that humans have ever sent into space—the *Voyagers* will still take about 30,000 years to reach the outer edge of the Oort Cloud (they should reach the inner edge in "only" about 300 years). The distances between asteroids and comets out there are so vast that it is highly improbable that either spacecraft will pass anywhere near any of them. Another 10,000 years later *Voyager 1* will pass only about 100,000 AU past the red dwarf star Gliese 445, which is now moving toward the sun and will by then be one of the closest stars to our solar system, just under four light-years away. Around the same time, *Voyager 2* will pass only about 111,000 AU from another red dwarf star, Ross 248, which will actually by then be the closest star to the sun in the sky. If there are inhabited planets around those stars—unlikely, given their small size and very faint output of energy as compared to our own sun, but still, who knows?—I wonder if anyone will notice these little emissaries zipping past from the star next door?

Indeed, Carl Sagan and his Golden Record colleagues speculated about whether it might be possible to avoid "the near certainty that left to themselves, neither Voyager spacecraft would ever plummet into the planet-rich interior of another solar system." Both they and I wonder if it might be possible to command one final "empty-the-tank" thruster firing, just before final communication with each *Voyager* is lost, to "redirect the spacecraft as closely as possible so that they will make a true encounter [with these stars]. If such a maneuver can be affected then some 60,000 years from now one or two tiny hurtling messengers from the strange distant planet Earth may penetrate into their planetary systems." If no one else does, I will try to remember to make this request to Suzy Dodd or whoever is running the *Voyager* Project in a decade or so, as the spacecraft power levels wind down. We have the fuel. Feel free to mention it to your congressperson.

It might also be interesting to see if we can *upload* images back onto *Voyager*'s tape recorders before we lose communication with them forever. While their Golden Records tell of their home world, there is nothing onboard each spacecraft that tells the stories of their magnificent adventures within their solar system. It's the same sentiment that motivates Jon Lomberg to pursue his "One Earth" *New Horizons* digital message project. "One thing I wish could have been on the *Voyager* Record," he told me, "which we are going to remedy with the *New Horizons* digital message, is that I wish we could have had something of 'here's what *Voyager* was and here's what Voyager found,' because it's one of the best things human beings have ever done. If they ever find *Voyager* they won't know about its mission. They won't know what it did, and that's sad." So I say let's try to upload the Earth-Moon portrait; the historic first close-up photos of Io's volcanoes and Europa and Ganymede's cracked icy shells; the smoggy haze of Titan; the enormous cliffs of Miranda; the strange cantaloupe and geyser terrain of Triton; the swirling storms of Jupiter, Saturn, and Neptune; the elegant, intricate ring systems of all four giant planets; the family portrait of our solar system. Let's arm our *Voyagers* with electronic postcards so that they can properly tell their tales, should any kind of intelligence ever find them.

INTELLIGENT BITS AROUND THE GALAXY

The idea of astronomers on other planets potentially noticing the *Voyagers,* or other signs of our civilization, is not as far-fetched as it used to be. In the past two decades, astronomers have discovered the first evidence of planets around other nearby stars that are like our sun. The first such planets were found using sensitive instruments on ground-based telescopes to search for tiny wobbles in a star's motion caused by the gravitational tug of planets orbiting around it. That technique, called the *radial velocity method,* is most sensitive to really big planets that are really close to their parent stars, causing a big wobble. And indeed, astronomers began finding dozens and dozens of these so-called hot Jupiters (named because they're Jupiter-sized but much hotter than our Jupiter because they are much closer to their parent stars) orbiting nearby stars. Some stars even have multiple close-in giant planets. Strange indeed, and nothing like our own solar system, where the giant planets are far away from their star. So is our solar system an oddball, and most stars have giant planets orbiting close-in? Or is our system typical, and the only reason we're finding so many hot Jupiters is the streetlight effect: we're finding what we're seeking only because we're looking where it's easiest to find exactly those things (like looking near the streetlight at night for your lost car keys, because that's where the light is). In this case, just because hot Jupiters are the easiest planets to find using the radial velocity method, that doesn't mean that they are the most common kind of planet out there.

Other methods were needed, to counteract the biases of the radial velocity method and to help find out what is a "typical" solar system in our galactic neighborhood. One promising idea, which had been advocated for decades by my former NASA Ames Research Center colleague Bill Borucki and others, was to build a camera that would patiently stare at the light from many stars and hope that some of them have planets that, occasionally, will pass in front of their star as seen from our perspective and dim that star's light by a tiny but measurable amount. This is called the *transit method*, because the planet passes across—or transits—its parent star. We can see this from time to time from Earth in our own solar system: both Mercury and Venus occasionally transit in front of the sun. Mercury last transited the sun in 2006 and will next transit in 2016.

Venus last transited the sun in 2004 and 2012 and won't transit the sun again from our perspective until the years 2117 and 2125. Transits are rare events, but they do happen if you're in the right place at the right time, and if you were to observe huge numbers of nearby stars, then even rare events like that should happen to some of those stars now and then.

That was exactly the philosophy that Bill Borucki had in mind when he and colleagues pitched the *Kepler* space telescope mission (named after the planetary-orbit discoverer Johannes Kepler) to NASA in 2001. Bill's idea was to launch a very sensitive camera and telescope—so sensitive that they could detect a 0.002 percent change in the light of a star corresponding to the dimming caused by transiting Earth-sized or smaller planets—and to literally stare at the same 150,000 stars or so for years to detect such transits. On the one hand, the *Kepler* mission has got to represent the most boring mission ever conducted, as it would orbit far beyond the Earth, and simply stare at the same region of space (a random patch of sky about as big as your fist held out at arm's length, in the northern constellations Cygnus and Lyra) over and over and over, radioing the same picture back to Earth again and again and again. But *some* of those stars in the picture should, statistically, show occasional transiting planets, NASA and planetary science colleagues realized, and by carefully tracking those statistics, we could figure out what percentage of stars have Earth-sized (or smaller) to Jupiter-sized (or larger) planets, and how far from their parent star these new worlds most commonly orbit. So, maybe in the end, *Kepler* could turn out to be one of the most exciting missions ever conducted.

Happily, that is what has come to pass. *Kepler* was launched in 2009, and within a few months of taking pictures, we began to see transits occurring from the "low-hanging fruit"—the easiest planets to find—the same kinds of Jupiter-sized worlds orbiting close to their parent star that the radial velocity method was so good at finding. But over the course of several years, as more images were taken and as the *Kepler* team got better at working with their data, smaller and smaller planets, orbiting farther from their parent star, were revealed. To date, *Kepler* has found more than 1,000 planets around nearby stars. The most exciting findings have been for a small fraction of those planets that are close to Earth in size and orbiting in the so-called habitable zone around their parent star. The habitable zone is the region around a star where an Earthlike planet with an Earthlike atmosphere could have stable liquid water on its surface and thus potentially support life as we know it. In our own solar system, the habitable zone runs from about the orbit of Venus to about the orbit of Mars. The surfaces of Venus and Mars themselves are not presently considered habitable because their atmospheres are significantly non-Earthlike (Venus's CO_2 atmosphere is hot and 100 times thicker than Earth's; Mars's CO_2 atmosphere is cold and 100 times thinner than Earth's). But they are still in roughly the right part of our solar system to have been habitable, if their atmospheric conditions were different in the distant past, for example. And Mars, as we now know from rovers and other missions sent there, could very well still have habitable zones lurking underground, where water, heat, and organic molecules might still all be found.

Earth is, of course, the Goldilocks planet—not too hot, not too cold, just right—and our planet defines what it means to be a habitable world in a star's habitable zone. A luckily placed alien astronomer on a nearby star might be able to see the Earth transit in front of the sun from their perspective. But they might not be able to tell the difference between a transiting Earth and the transiting of a sunspot, so they'd want to see it happen again. So they'd have to wait for a year—there it would go again! No good scientist would be able to convince skeptical colleagues that this wasn't just coincidence, however, so that alien scientist would have to predict the next transit . . . wait another year . . . and voilà! If a slight drop in the star's light is seen again right when predicted, they know they've bagged a real planet. That same philosophy is used by the *Kepler* team, which is why they often need three or four years of data to find proof of the discovery of Earth-sized planets in the habitable zones around nearby sunlike stars.

I had a short involvement with the *Kepler* Project and the search for extrasolar planets in the early 1990s, after I graduated from the University of Hawaii and moved on to a postdoc position at the NASA Ames Research Center at Moffett Field, just north of San Jose, California. Bill Borucki gave a lunchtime talk about his (then) crazy-sounding idea for a space-based mission to use transits to find Earthlike worlds around other stars. Bill is a soft-spoken guy but from the beginning has been one of the world's most passionate advocates of the possibilities for discovery using the transit method. I asked if there was some small way I could get involved, some minor project I could help to advance in some way. He mentioned that not many people had been thinking yet about what kinds of transits could be detected around binary star systems. Since most stars in the galaxy that are similar to our sun are members of binary (or more) star systems, it seemed like a relevant topic to explore. I reminded myself of Kepler's laws of motion, wrote some computer programs to simulate stars in orbit around each other, and added some computerized planets in different places in the system. I found very different, but distinctive, transit signatures from planets orbiting both stars, and planets orbiting just one of the binary companions.

We wrote up some short descriptions of our results as abstracts for a few conferences, including a very unconventional and eclectic workshop called the Bioastronomy Symposium: Progress in the Search for Extraterrestrial Life, held in Santa Cruz in 1993 and attended by a wild mix of scientists, artists, philosophers, musicians, science-fiction writers, biologists, cosmologists, and, it seemed, perhaps even some actual clinically insane people, all thinking about the search for habitable planets and life beyond Earth. Unfortunately, I got bogged down in other projects and never wrote up our results in a peer-reviewed journal. After Bill finally got his *Kepler* mission approved and he and other colleagues finally did find evidence for planets around binary star systems—the first was reported in a paper published in *Science* magazine in September 2011—I was excited that the kinds of solar systems that we had dreamed about back then actually turned out to exist, and some had the kinds of transit signals that we had predicted in our long-ago study.

Some of the most interesting kinds of planets that *Kepler* is discovering are dubbed super-Earths or mini-Neptunes. Approaching or comparable in size to Neptune, these worlds are either ice giants like Uranus and Neptune or much larger, rocky planets like Earth that have become large enough that their gravity can help them hold on to thick atmospheres as they are forming. It turns out that these kinds of planets—and ice giants in particular—are, so far, the most common kinds of worlds seen around other nearby sunlike stars in the *Kepler* transit images. How fortunate, then, that *Voyager 2* was able to characterize in detail two such planets right here in our own solar system. Indeed, the discovery from *Voyager 2* data that Uranus and Neptune are not gas giants but are instead a separate, new class of planets called ice giants has turned out to be fundamental to understanding the figurative zoo of planetary types being found by the *Kepler* team. And importantly, the realization from *Voyager* and other missions that many of the *moons* of the giant planets are worthy of consideration as potential habitable worlds should be a warning to astronomers to not consider the idea of a close-to-the-star habitable zone as a firm and fixed constraint on where, in other solar systems, we might need to go looking for life.

THE END?

What does the future hold for the *Voyagers* themselves, the robotic harbingers of our new Interstellar Age? Most experts predict that by the mid- to late-2020s or so, their slowly decaying plutonium-238 power supplies will not be able to produce enough power to keep them heated and in contact with the Earth. They will fall into a long, permanent, silent deep freeze. The cold and vacuum of space will silence their hearts but will not cease their traveling on to the stars. Micrometeorites and high-energy cosmic rays will occasionally strike them, creating tiny pits or other local-scale metallurgic damage probably too small to notice for any particular event, but certainly more significant when summed over thousands and millions of years. The *Voyagers'* Golden Records are better protected than many other parts of the spacecraft, nestled inside their gold-plated aluminum cases on each vessel. The records' outer-facing surface is predicted to be pitted by micrometeorites over less than 2 percent of its surface by the time the ships are a full light-year away. Because of the lower density of particles in interstellar space, it should take an additional 5,000 light-years of travel, or about 100 million years, for the outer surface of the records to accumulate an additional 2 percent damage. During this time the inner surface of the records, protected even more from the "elements" of space weathering, will remain essentially pristine. Thus, barring a highly unlikely catastrophe such as a chance impact with a rogue asteroid or comet, or an extremely low-probability capture and subsequent burn-up of the spacecraft by a nearby star, there is nothing that should destroy or seriously degrade the physical integrity of the *Voyagers* and their precious messages from Earth. Will they, then, continue to ply the stars forever?

I am not so sure, because forever is an awfully long time. I believe that someday, in the far future, it will be humans, rather than extraterrestrials, who might have the highest probability of catching up to the *Voyagers*. It seems inevitable, and I can imagine many possible scenarios for that encounter. Imagine, for example, that 50,000 years from now an enormous generational starship, launched by a consortium of privately funded gajillionaires seeking new adventures and new opportunities, heads off for the solar system around the star Gliese 445. That particular star will be only about 3 light-years from the sun at the time, and astronomers have discovered an Earthlike moon orbiting one of its gas giant planets. As they near the star, the settlers decide to search for the ancient *Voyager 1* probe, which is thought to be traveling in the same general vicinity. They find and gently scoop up the frail, spindly spacecraft and lovingly set it up as a monument to the beginning of the Interstellar Age in their ship's main atrium. Plans start to be formulated about a more permanent monument to be built around the spacecraft when they arrive at their new home world.

In a little less than 300,000 years, *Voyager 2* will pass about 270,000 AU (about 4.3 light-years) from the famous young, hot, blue star called Sirius. Sirius is famous partly because it is the brightest star in the sky, aside from the sun, and partly because ancient civilizations, such as the Egyptians, Greeks, and Polynesians used Sirius for timekeeping and navigation. *Voyager 2* will be half as far away from Sirius as we are now, and so what many of us call the Dog Star (the heart of the constellation Canis Major) will be four times brighter to the spacecraft. An impressive sight that would be, if we could somehow plan to turn the cameras back on in the year 298,015. Beyond then, it's hard to know exactly what the *Voyagers* will pass and when, because of uncertainties in the relative motions of the sun and the nearby stars. Certainly over future millions of years the spacecraft will make other encounters within several light-years of other nearby stars. The thrusters may fire one more time and propel one or the other of our *Voyagers* into another solar system, or maybe not. Maybe there will be some form of intelligent life in that solar system. Who knows.

More generally, though, both spacecraft are destined to follow long, relatively circular, 250-million-year-long orbits around the center of the Milky Way galaxy, just like the sun and many other nearby stars. The *Voyagers* are interstellar travelers, but they are not intergalactic travelers—they would have to have been accelerated about fifteen times faster than they're currently going, or about 1 million miles per hour, to escape the gravity of the Milky Way. It's nice to know that even in the far future, though long dormant, the *Voyagers* will still be making graceful trips around the galactic center with us.

Still, *Voyager 1* will continue to slowly travel northward and *Voyager 2* southward, relative to the sun and the surrounding stars. Over time—enormous spans of time, as the gravity of passing stars and interacting galaxies jostles them as well as the stars in our galaxy—I imagine that the *Voyagers* will slowly rise out of the plane of our Milky Way, rising, rising, ever higher above the surrounding disk of stars and gas and dust, as they once rose above the plane of their home solar system. If our far-distant descendants remember them, then our patience, perseverance, and persistence could be rewarded with perspective when our species—whatever it has become—does, ultimately, follow after them. The *Voyagers* will be long dormant when we catch them, but they will once again make our spirits soar as we gaze upon these most ancient of human artifacts, and then turn around and look back. I have no idea if they'll still call it a selfie then, but regardless of what it's called, the view of our home galaxy, from the outside, will be glorious to behold.

My ASU colleague and *Voyager* historian Stephen Pyne has noted, poignantly, "Even as they are celebrated for racing forward—to the outer planets, to the heliopause, to interstellar space—many of their most dazzling discoveries were the offshoot of staring back at what they passed in their sling-shot fly-bys. Their trajectory is a triangulation of future and past, or what might be recalibrated as expectation and meditation. The *Voyagers* were special when they launched. They have become more so thanks to their longevity, the breadth of their discoveries, the cultural payload they carried, and the sheer audacity of their quest."

"That's the thing about this mission," offered Ed Stone, reflecting on *Voyager*'s transition from a planetary to an interstellar mission, "there really hasn't been an end. All of these encounters and events, are in some sense 'ends,' but they're really not the end. *That's* really the wonderful thing about *Voyager.*"

Postscript: NewSpace

HUMANS ARE CURRENTLY controlling about thirty different spacecraft that, together, make up an impressive robotic armada sent out to explore our solar system and beyond. At times we strain to see beyond the turmoil and the swift pace of modern life. We forget to take a moment to look outward and perceive our place in time. We are all living—right now—in an amazing Golden Age of Exploration, of our planet and of our solar system. And if we look closely, in our mind's eye, we can see the *Voyagers* quietly ushering us to and across the threshold of the Interstellar Age.

How should the future of space exploration unfold? Will NASA and other government space agencies always lead the way? Will upstart private space-related companies (such as SpaceX, Virgin Galactic, Sierra Nevada, and dozens of others) dive into the robotic space-exploration game, and if so, why? For mineral resources? For fame and glory? To help protect the Earth from rogue asteroids or comets? To settle other worlds? Or will those companies instead focus mostly on providing services like lower-cost rocket launches, adventure-tourism experiences, space-station refueling/resupply, or satellite repair? Those questions are at the forefront of the emerging space industry sector often called NewSpace. Investors large and small are trying to predict which of these businesses, and which of these questions, will drive the future of space-related business and research. And perhaps unbeknown to most people, governments are playing important roles in the increasing privatization of space-related activities. NASA, for example, in their Commercial Orbital Transportation Services and other Commercial Crew & Cargo Programs, has doled out nearly $2.5 billion of taxpayer funding over the past five years or so to spur the development of many of these "private" initiatives. In many ways, the government is playing a similar role today in the creation of a privately run, civil space program that it played in the early to mid-twentieth century in civil aviation. In the 1920s, for example, the US government was the biggest and most reliable customer for the nascent airline industry, paying out sweet contracts for the delivery of airmail to then-upstart companies with names like TWA, Northwest, and United. Which private space companies being seeded by tax dollars today will emerge as household names and NewSpace industry giants of the twenty-first century?

While missions like *Voyager* were science- and exploration-driven, many NewSpace (and traditional "OldSpace") companies are of course bottom-line- and profit-driven. Still, there is great potential for collaboration and cross-fertilization. By analogy, fruitful partnerships have emerged in recent decades between leading environmental and ecological preservation groups and organizations and some parts of the worldwide tourism industry. Nonspecialist individuals and families can now take vacations that also support scientific research projects related to the ecology, archaeology, sociology (and so on) of their destinations. I believe that a similar model could be highly effective for space-related tourism. What many space scientists want most for their research is *access* to the space environment—be it experiments in low gravity, or new measurements from orbital or landed/roving platforms—and an *interested, excited audience* (preferably decision makers at funding agencies, but really anyone genuinely interested is valued) with whom to share their results. If that access is provided by private NewSpace companies, and part of the price is that researchers have to be tour guides and teachers for the paying public sharing the ride, so be it. It's a model that could work.

I like to imagine an entirely new branch of nerdy but potentially lucrative adventure tourism that could easily, in the not-too-distant future, be built around what I call "manufactured astronomical events." We've all seen photos of the magnificent splendor of a total solar eclipse, for instance, but very few people have actually *seen* a total solar eclipse because they occur only about once every three hundred years in any particular city or region on Earth. But that's because of the particular geometry of the sun, Earth, and moon for *Earthbound* observers. If we were to take a spaceship to the right places in space, it would be easy to fly the ship through the shadows of the Earth or the moon and re-create the same kind of eclipse experience for those aboard the ship. As another example, there was a lot of hubbub a few years back about people viewing the last transit of Venus across the disk of the sun until the year 2117. Not necessarily so! Take a ship full of astronomical adventure tourists to the right place in space and at the right time to watch, and—voilà!—there's a Venus transit as good as any that you'd see from Earth. Many other kinds of "manufactured" celestial events will be possible to create once access to near-Earth (and lunar) space becomes more routine. We could experience Earth-and-moon solar eclipses, transits of Mercury or other planets, flights through active comet tails, flybys and landings on near-Earth asteroids, perhaps even visits to *Voyager* and other ancient spacecraft. Such excursions may never become as routine as airline travel is today, but I believe that the trend will surely be toward safer, more affordable, and more personally meaningful access to space for regular citizens.

Notes and Further Reading

Direct quotes in the book come from a series of interviews that I conducted during 2013 and early 2014 with many *Voyager* friends and colleagues that I've had the privilege of talking with and/or working with. I recorded and transcribed the interviews, and confirmed the use of the quotes with each interviewee. I have listed these folks among the Acknowledgments because I am truly grateful for their time and patience!

Chapter 1. Voyagers

19 the privilege of walking on another world: To be the hit of the party sometime (at least, the kinds of parties I go to . . .), memorize and then recite all twelve of their names! In order of their missions (*Apollos 11–12, 14–17*): Neil Armstrong and Edwin "Buzz" Aldrin; Pete Conrad and Alan Bean; Alan Shepard and Edgar Mitchell; James Irwin and David Scott; Charles Duke and John Young; and Harrison "Jack" Schmitt and Gene Cernan.

19 so cute with their long necks and bulging eyes!: Does "cuteness" matter in space exploration? See planetary scientist Melissa Rice's thoughts on the topic in her Planetary Society blog entry "In Memory of *Spirit*, and Why Cuteness Matters" at planetary.org/blogs/guest-blogs/3065.html.

21 **largest public-membership space-advocacy organization:** You can learn more about the history and vision of The Planetary Society at planetary.org. I can't help but sing the accolades of this merry band of fellow space explorers, as I happen to be the society's president!

22 **some in Congress have asked (*really*):** For an example, see NASA historian Stephen J. Garber's article "Searching for Good Science: The Cancellation of NASA's SETI Program," *Journal of British Interplanetary Society* 52 (1999): 3–12 (online at history.nasa.gov/garber.pdf).

22 **Why should American taxpayers support NASA?:** Wikipedia has a fairly comprehensive entry on the history of the NASA budget, with links to more information, at en.wikipedia.org/wiki/Budget_of_NASA.

25 **inspiration is priceless during tough times:** Watch and read Neil deGrasse Tyson's passionate 2012 testimony to the US Senate's Commerce, Science, and Transportation Committee on Neil's own website, at haydenplanetarium.org/tyson/read/2012/03/07/past-present-and-future -of-nasa-us-senate-testimony.

39 **complications of a stroke, passed away in late 2005:** A nice "In Memoriam" piece written by several of Ed Danielson's professional colleagues can be found in the planetary science journal *Icarus* 194 (2008): 399–400 (online at dx.doi.org/10.1016/j.icarus.2007.12.007).

Chapter 2. Gravity Assist

43 **"the basic ideas behind gravity assist . . .":** David W. Swift, *Voyager Tales: Personal Views of the Grand Tour* (Reston, VA: American Institute of Aeronautics and Astronautics, 1997), page 63.

44 **pass each planet with the shortest possible trip time:** Ibid., page 64.

45 **The next opportunity would not appear:** Ibid.

46 **"Many openly scoffed at the idea,":** Ibid., page 66.

47 **"Many myths have arisen . . .":** Ibid., page 69.

47 **"Those at JPL who brought everything together . . .":** Ibid.

47 **then two in 1979 to fly by Jupiter, Uranus, and Neptune:** *1970 Annual Report*, Jet Propulsion Laboratory, California Institute of Technology, Pasadena, CA, page 8 (online at http://www.jpl.nasa.gov/report/1970.pdf).

48 **"They told us, 'If you guys . . .":** Douglas Smith, "The Other Side," *Engineering & Science Magazine*, California Institute of Technology, Pasadena, CA, Winter 2013, 10–13.

50 **built around a basic chassis called a *bus*:** Dave Doody, *Deep Space Craft: An Overview of Interplanetary Flight* (New York: Springer/Praxis, 2009), page 143.

61 **significantly changed from the original *Mariner* configuration:** Andrew J. Butrica, "Voyager: The Grand Tour of Big Science," in *From Engineering Science to Big Science*, ed. Pamela E. Mack, NASA History Office Special Publication 4219 (Washington, DC: National Aeronautics and Space Administration, 1998), 251–76 (online at history.nasa.gov /SP-4219/Contents.html).

68 **complete the Great Pyramid at Giza for King Cheops:** C. Kohlhase, ed., *The Voyager Neptune Travel Guide*, JPL Publication 89-24 (Pasadena, CA: Jet Propulsion Laboratory, California Institute of Technology, 1989), page 135 (online at babel.hathitrust.org/cgi/pt?id=uiug.3011205 6430637).

Chapter 3. Message in a Bottle

73 **"We step out of our Solar System into the universe . . .":** Carl Sagan, F. D. Drake, Ann Druyan, Timothy Ferris, Jon Lomberg, and Linda Salzman Sagan, *Murmurs of Earth: The Voyager Interstellar Record* (New York: Random House, 1978), page 26. See also an online index of the "Scenes from Earth" photo collection at voyager.jpl.nasa.gov/spacecraft /scenes.html and an online index of the "Music of Earth" music collection at voyager.jpl.nasa.gov/spacecraft/music.html.

75 **cast out forever into interstellar space:** For details on the origin and content of the *Pioneer* plaques, see Carl Sagan, Linda Salzman Sagan, and Frank Drake, "A Message from Earth," *Science* 175, no. 4024 (1972): 881–84 (online at sciencemag.org/content/175/4024/881 .short).

76 **"carry some indication of the locale . . .":** Ibid., page 881.

76 **"Pioneer 10 and any etched metal . . .":** Ibid.

77 **"the message can be improved . . .":** Ibid., page 883.

79 **"If we don't send things we passionately care for . . .":** Sagan et al., *Murmurs of Earth*, page 254.

80 **"Hello to everyone . . .":** Ibid., page 143.

81 **a variety of nearby stars in 1999 and 2003:** The "Cosmic Call" refers to two sets of messages sent to nearby stars from the RT-70 radio telescope facility in Yevpatoria, Crimea, in 1999 and 2003. See en.wikipedia.org /wiki/Cosmic_Call for more details.

81 **"which didn't turn out very well . . .":** Stephen Hawking, *Into the Universe with Stephen Hawking*, Television Series, Episode 1: "Aliens," Discovery Channel, 2010.

84 **Martian sundials:** Woody Sullivan and Jim Bell, "The MarsDial: A Sundial for the Red Planet," *The Planetary Report* (January/February 2004): 6–11.

89 **"It's wise to try . . .":** Michael D. Lemonick, "Life beyond Earth," *National Geographic*, July 2014, page 44.

94 **Photos and Diagrams on the *Voyager* Golden Record:** Sagan et al., *Murmurs of Earth*, pages 71–122.

96 **Music on the *Voyager* Golden Record:** Ibid., pages 161–209.

98 **140 countries signed online petitions:** Jon Lomberg's "One Earth: New Horizons Message Project" website can be found at oneearthmessage.org.

99 **the One Earth: New Horizons Message Project:** Ibid.

Chapter 4. New Worlds among the King's Court

105 **the right amount to swing the probe on to Saturn:** Charley Kohlhase's ongoing memoirs, "The Complete Rocket Scientist," which includes fascinating technical details about how he and the *Voyager* team designed the Jupiter-Saturn-Titan and Jupiter-Saturn-Uranus-Neptune trajectories of *Voyagers 1* and *2*, is online at charleysorbit.com/completerocketscientist/lifebook1.php.

106 **the equivalent of about 1 foot per *trillion years*:** Kohlhase, *Voyager Neptune*, page 139.

113 **what those in the business call a *3-axis stabilized* spacecraft:** For all kinds of wonderful detail, diagrams, and descriptions of *Voyager*'s many spacecraft systems and instruments, see "The *Voyager* Spacecraft," by former project manager Raymond L. Heacock, published in the *Proceedings of the Institution of Mechanical Engineers* 194, no. 28 (1980): 211–24 (online at stickings90.webspace.virginmedia.com/voyager.pdf).

115 **a team of three celestial mechanics experts:** S. J. Peale, P. Cassen, and R. T. Reynolds, "Melting of Io by Tidal Dissipation," *Science* 203, no. 4383 (1979): 892–94 (online at http://dx.doi.org/doi:10.1126/science.203.4383.892).

115 **"*Voyager* images of Io may reveal . . .":** Ibid., page 894.

117 **the Io images taken for navigation purposes:** The "discovery photo" where Linda Morabito and colleagues first noticed the volcanic plumes of Io is online at photojournal.jpl.nasa.gov/catalog/PIA00379.

118   **were eruption plumes from active volcanoes on Io:** For a full, detailed, first-person account of the events leading up to the historic discovery of Io's active volcanic plumes, see Linda Morabito, *Discovery of Volcanic Activity on Io*, archived online at arxiv.org/pdf/1211.2554; for more stories and details about the study and discovery of volcanoes on other worlds since then, see also Rosaly Lopes and Tracy Gregg, eds., *Volcanic Worlds: Exploring the Solar System's Volcanoes* (New York: Springer-Praxis Books, 2004).

125   **I can live to see that exploration pay off:** For a great update on the status of the search for life in extreme environments like Europa, as well as upcoming plans for Europa exploration, see Michael D. Lemonick's "Life beyond Earth" in the July 2014 issue of *National Geographic* magazine.

130   **"By far the simplest explanation . . .":** Lorenz Roth's December 2013 press release describing the Hubble Space Telescope's potential discovery of plumes of water vapor coming from Europa can be found online at http://www.nasa.gov/content/goddard/hubble-europa-water-vapor.

131   **"Decadal Survey of Planetary Science":** The recent 2011 National Academy of Sciences Decadal Survey of Planetary Science, *Visions and Voyages for Planetary Science in the Decade 2013–2022* can be found online from the National Academies Press at nap.edu/openbook.php?record_id=13117.

Chapter 5. Drama within the Rings

135   **inner and outer parts orbiting at different speeds:** For a fascinating and entertaining account of Maxwell's work on Saturn's rings, as well as his fundamental contributions to the physics of electricity and magnetism, see Basil Mahon, *The Man Who Changed Everything: The Life of James Clerk Maxwell* (Hoboken, NJ: Wiley, 2003).

136   **perspectives to come from the *Voyagers* that would follow:** Amateur astronomer and planetary image processor Ted Stryk has compiled a nice collection of *Pioneer 11* "greatest hits" images of Saturn online at strykfoto.org/pioneersaturn.htm.

139   **perhaps some other complex hydrocarbons:** This and other early pioneering planetary spectroscopic discoveries were made by the Dutch-American astronomer Gerard P. Kuiper, who is widely regarded as one of the founding fathers of modern planetary science. There's a nice Wikipedia biography of Kuiper online at en.wikipedia.org/wiki/Gerard_Kuiper.

139   **could have led to the formation of life on Earth:** Wikipedia's entry on the Miller-Urey experiments at en.wikipedia.org/wiki/Miller-Urey _experiment is a great starting point for learning more about these famous

early efforts at understanding the possible origins of life on Earth and other habitable worlds.

141 *Voyager*'s **cameras were blind to the surface itself:** Uncovering those secrets, including discovering the hoped-for seas of ethane and methane, would have to wait more than twenty-five years, when the *Cassini* Saturn orbiter, armed with cloud-penetrating radar inspired by *Voyager*'s discoveries, would finally map the fascinating geology and hydrology of Titan and when the ESA *Huygens* probe would get near-surface images just before landing. My planetary science colleagues Ralph Lorenz and Christophe Sotin wrote an exciting and accessible article about our current knowledge of enigmatic Titan, "The Moon That Would Be a Planet," in the March 2010 issue of *Scientific American* magazine.

142 **at the time the solar system's most distant known planet:** For some more details about whether *Voyager* could have encountered Pluto, and other *Voyager* "Frequently Asked Questions," see the *Voyager* Project's official FAQ at voyager.jpl.nasa.gov/faq.html.

145 **"fresh" ice in the process:** It's a debate that even higher-resolution *Cassini* Saturn orbiter images and other data have yet to resolve. *Cassini* camera team leader Carolyn Porco, who was also heavily involved in *Voyager* imaging of Saturn's rings earlier in her career, has been careful to note that "there's a bunch of caveats in all of this. Very little in this area is definite. Each part of the rings may have a different age." (Porco, as quoted in Richard A. Kerr, "Saturn's Rings Look Ancient Again," *Science* 319 (2008): 21.) It seems that, once again, we find that we really do have to go back. . . .

145 **rocky lava flow would on Earth:** A great introduction to cryovolcanism can be found in Rosaly Lopes and Michael Carrol's book *Alien Volcanoes* (Baltimore: Johns Hopkins University Press, 2008).

152 **the team was confident that everything would go as planned:** A day-by-day, diary-like account of the details of the *Voyager* Saturn encounters was published by *Voyager* imaging team member David Morrison in *Voyages to Saturn*, NASA Special Publication 451 (Washington, DC: National Aeronautics and Space Administration, 1998) (online at babel .hathitrust.org/cgi/pt?id=uiug.30112012462427). Pages 185 to 189 of that book contain a full list of the principal investigators and co-investigators on the *Voyager* science team and the leaders of the *Voyager* management teams.

154 **"a million times the normal energy level . . .":** Fred Scarf, quoted in Morrison, *Voyages to Saturn*, page 123.

154 **almost minute-by-minute account of the Saturn flybys:** Morrison, *Voyages to Saturn*, page 123.

154 **"The quantity of such impacts . . .":** Ibid.

156 **in his diary of events for August 28, 1981:** Ibid., page 131.

157 *Cassini* **orbiter on a "suicide mission" to Saturn:** Richard C. Hoagland's original June 30, 2004, article (and his quote) about this can be found online at http://www.enterprisemission.com/_articles/06-30-2004_Cassini/IsNASA SendingtheCassiniMissiontoitsDoom.htm.

157 **"a sense of gloom . . .":** Ibid., page 119.

Chapter 6. Bull's-Eye at a Tilted World

165 **a planetary path, well beyond the orbit of Saturn:** I recount a lot of the interesting history of William Herschel's and other pioneering discoveries in astronomy in *The Space Book: 250 Milestones in the History of Space and Astronomy* (New York: Sterling, 2013).

166 **important work cataloguing faint stars:** A nice article about Caroline Herschel's life and achievements, including her support of her famous brother's astronomical discoveries, appears in J. Donald Fernie's "The Inimitable Caroline" in the November/December 2007 issue of *American Scientist* (online at americanscientist.org/issues/pub/the-inimitable-caroline).

170 **new tilt for a newly formed (potentially merged) planet:** For a short summary of some of the latest ideas about the strange tilt of Uranus, see John Matson's October 7, 2011, article "Double Impact: Did 2 Giant Collisions Turn Uranus on Its Side?" in *Scientific American* (online at scientificamerican.com/article/uranus-axial-tilt-obliquity).

175 **"That all of this worked so well . . .":** Edward C. Stone and Ellis D. Miner, "The *Voyager 2* Encounter with the Uranian System," *Science* 233 (1986): 39–43.

180 **ten new, smaller moons:** Wikipedia's "Moons of Uranus" page, at en.wikipedia.org/wiki/Moons_of_Uranus, is a great resource for history, facts, and additional research links on the twenty-seven presently known moons of the seventh planet.

183 **seems to be a leading hypothesis for what happened:** Space.com editor and space history author Andy Chaikin posted an excellent article titled "Birth of Uranus' Provocative Moon Still Puzzles Scientists" on October 16, 2001, summarizing the post-*Voyager* state of confusion regarding the history of Miranda, online at archive.today/6VTxV.

184 **astronomers led by the late Jim Elliot of MIT:** For a firsthand account of the discovery of the Uranian rings, see Jim Elliot and Richard Kerr, *Rings: Discoveries from Galileo to Voyager* (Cambridge, MA: MIT Press, 1987).

188 **a peer-reviewed scientific journal article:** My research paper, coauthored with my PhD dissertation advisor, T. B. McCord, is titled "A Search for Spectral Units on the Uranian Satellites Using Color Ratio Images" and was published in the *Proceedings of Lunar and Planetary Science* 21 (1991): 473–89 (online at http://adsabs.harvard.edu/abs/1991LPSC...21.473B).

188 **more complex and dynamic weather:** For a great summary of the Hubble Space Telescope's history of imaging of Uranus, check out hubblesite .org/newscenter/archive/releases/solar-system/uranus.

Chapter 7. Last of the Ice Giants

193 **nothing less than solar-system glory:** In *The Planet Neptune: An Historical Survey before Voyager* (New York: Wiley, 1996), the late Sir Patrick Moore, British astronomer and science popularizer, provides additional stories and details about the history of the discovery of Neptune.

199 **taxpayers of an entire nation:** For an interesting historical perspective on *Voyager* within the broader context of exploration over the centuries, see Stephen Pyne, *Voyager: Seeking Newer Worlds in the Third Great Age of Discovery* (New York: Viking, 2010).

202 **were found to be roughly ten times smaller:** Ibid., page 140.

207 **changes in the atmosphere of Neptune over time:** For a great summary of the Hubble Space Telescope's history of imaging of Neptune, check out hubblesite.org/newscenter/archive/releases/solar-system/neptune.

209 **entirely new and unanticipated class of planet:** For a fun and educational conversation about ice giants, check out Planetary Society Weekly Hangout blogger Emily Lakdawalla's April 11, 2013, interview with *Voyager* imaging team member and Planetary Society vice president Heidi Hammel, online at planetary.org/blogs/emily-lakdawalla/2013 /hangout-20130409-heidi-hammel.html.

212 **the entire solar system by nearly 70 percent:** Just as for all the other planets with moons, Wikipedia is a great resource to learn more about the fourteen moons presently known around Neptune. See en.wikipedia .org/wiki/Moons_of_Neptune.

217 **the low gravity and atmospheric pressure:** For all the gory details, see "Triton's Geyser-like Plumes: Discovery and Basic Characterization," by Larry Soderblom and eight other *Voyager* team colleagues, in the October 19, 1990, issue of *Science* magazine (vol. 250, no. 4979, 410–15).

217 **KBOs, as they are now known, have been discovered:** The International Astronomical Union's Minor Planet Center, operated by Harvard's Smithsonian Astrophysical Observatory, keeps up-to-date lists and graphical plots of the orbits and positions of all the more than 650,000 presently known asteroids and comets in the solar system, including the Kuiper Belt Objects. The lists are online at minorplanetcenter.net/iau /lists/MPLists.html; to see the KBO list, click on "Transneptunian Objects," the more generic name for KBOs.

218 *New Horizons* **spacecraft flies by Pluto in July of 2015:** Visit the *New Horizons* mission website for more details at pluto.jhuapl.edu.

Chapter 8. Five Billion People per Pixel

224 **human beings who have lived before us:** The Population Reference Bureau has an excellent online article explaining their estimate for the total number of people who have ever lived on Earth, posted at prb.org/Publications/Articles/2002/HowManyPeopleHaveEverLivedon Earth.aspx.

226 **the White Sands Missile Range in New Mexico:** Tony Reichhardt, "The First Photo from Space," *Air & Space* magazine, November 2006 (online at airspacemag.com/space/the-first-photo-from-space-13721411/?no-ist).

226 **after the first Earth-orbiting satellites were launched:** For more photos and details about these early photos of the Earth from space, see, for example, (a) http://www.nasa.gov/centers/langley/home/Road2Apollo -11_prt.htm; (b) space.com/12707-earth-photo-moon-nasa-lunar-orbiter -1-anniversary.html; (c) moonviews.com/lunar-orbiter-1-i-or-a; and (d) moonviews.com/2013/05/how-life-magazine-revealed-earthrise-in-1966 .html.

228 **"Oh my God. Look at that picture . . .":** Transcripts of conversations and events from the *Apollo 8* mission can be found in David Woods and Frank O'Brien, "The *Apollo 8* Flight Journal," available online from the NASA History Division at http://www.history.nasa.gov/ap08fj.

232 **"The point of such a picture . . .":** Carl Sagan, "A Pale, Blue Dot," *Parade Magazine,* September 9, 1990, page 52.

236 **"a pale blue dot . . .":** Ibid.

237 **"It has been said that astronomy is . . .":** Carl Sagan, *Pale Blue Dot: A Vision of the Human Future in Space* (New York: Random House, 1994), pages 8–9.

237 **selfies by subsequent planetary exploration missions:** The Planetary Society, the world's largest nonprofit public space-advocacy organization, hosts a delightful online collection of photographs of the Earth from space at planetary.org/explore/space-topics/earth/pics-of-earth-by -planetary-spacecraft.html.

237 *Cassini* **orbiter was passing through Saturn's shadow:** Ibid.

238 **"After much work, the mosaic . . .":** Details and photos from the *Cassini* mission's "The Day the Earth Smiled" photo event can be found online at photojournal.jpl.nasa.gov/catalog/PIA17171 and on *Cassini* imaging team leader Carolyn Porco's Facebook page at facebook.com /carolynporco.

238 **inspirational appeal as the *Pale Blue Dot*:** For lots of stories and beautiful examples of some of the most spectacular photos from the *Spirit* and *Opportunity* rover missions, see my book *Postcards from Mars* (New York: Dutton, 2006).

239 **But the next day when the images were beamed back:** For some examples of the photos of Earth taken from the surface of Mars, see photojournal.jpl.nasa.gov/catalog/PIA05547 and also pancam.sese.asu .edu/pancam_instrument/projects_3.html.

Chapter 9. The Edge of Interstellar Space

242 **their "planet" status:** For lots more background and detail on the controversy over the demotion of Pluto, see Neil deGrasse Tyson, *The Pluto Files* (New York: W. W. Norton, 2009); Mike Brown, *How I Killed Pluto and Why It Had It Coming* (New York: Spiegel & Grau, 2012); and the details about the IAU's decision to demote Pluto to dwarf planet status, online at iau.org/public/themes/pluto.

244 **maybe 50,000 years ago or more:** See astronomer and science evangelist Phil Plait's *Bad Astronomy* blog entry titled "The Long Climb from the Sun's Core" at badastronomy.com/bitesize/solar_system/sun.html for information on how long it takes photons to escape the sun's core.

249 **are also on escape trajectories out of the solar system:** For live updates on the speeds and distances of the five spacecraft that humans have launched on escape trajectories from our solar system, see heavens -above.com/SolarEscape.aspx.

253 **"to extend the NASA exploration of the solar system . . .":** The *Voyager* Project at JPL hosts an official website describing the goals and achievements of the *Voyager* Interstellar Mission at voyager.jpl.nasa.gov /mission/interstellar.html.

255 **only 0.0000000000000001 watts, or barely a flea's whisper:** Kohlhase, *Voyager Neptune*, page 136.

258 **"I feel extremely fortunate . . .":** For more about Suzy Dodd, and a list of the nine previous *Voyager* project managers before her, see voyager.jpl .nasa.gov/news/dodd_proj_manager.html.

263 **"porous, multi-layered structure threaded by . . .":** M. Swisdak, J. F. Drake, and M. Opher, "A Porous, Layered Heliopause," *Astrophysical Journal* 774, L8 (2013): 1 (online at iopscience.iop.org/2041-8205/774 /1/L8/pdf/apjl_774_1_8.pdf).

264 **"We think we are outside . . .":** Marc Swisdak interviewed in Richard Kerr, "It's Official—*Voyager* Has Left the Solar System," *Science* 341 (September 2013): 1158–159.

265 **"When we saw that, it took . . .":** Don Gurnett quoted in Ibid., page 1159.

265–66 **"Now that we have new, key data . . .":** NASA's official September 12, 2013, press release, and Ed Stone's comments, in "*Voyager* Reaches Interstellar Space" (online at science.nasa.gov/science-news/science -at-nasa/2013/12sep_voyager1).

266 **"I don't think it's a certainty":** McComas quoted in *Science* 341 (September 2013): 1159.

266 **"We have not crossed the heliopause.":** George Gloeckler quoted in Ibid.

266 **particles and magnetic fields within the heliosphere:** L. A. Fisk and G. Gloeckler, "The Global Configuration of the Heliosheath Inferred from Recent *Voyager 1* Observations," *Astrophysical Journal* 776 (2013): 79 (online at iopscience.iop.org/0004-637X/776/2/79/pdf/apj_776_2_79.pdf).

Chapter 10. Other Stars, Other Planets, Other Life

275 **"One-hundred-six elements":** Glenn T. Seaborg won the Nobel Prize in Chemistry in 1951 for his work on the transuranian elements. See the Prize website's official biography of him for more details on his life and illustrious career at nobelprize.org/nobel_prizes/chemistry /laureates/1951/seaborg-bio.html.

277 **spacecraft's lifetimes beyond the mid-2020s:** The JPL *Voyager* Project's official website for tracking power conservation strategies and limitations is at voyager.jpl.nasa.gov/spacecraft/spacecraftlife.html.

278 **only slightly above the plane of the planets:** See heavens-above.com /SolarEscape.aspx.

279 **modern-day spacecraft forensics:** For more details, see The Planetary Society's director of projects Bruce Betts's April 19, 2012, blog post "Pioneer Anomaly Solved!" at planetary.org/blogs/bruce-betts/3459.html.

281 **just under four light-years away:** For information about *Voyager 1*'s predicted encounter with Gliese 445, see en.wikipedia.org/wiki /Gliese_445, and for information about *Voyager 2*'s predicted encounter with Ross 248, see en.wikipedia.org/wiki/Ross_248.

281 "redirect the spacecraft as closely as possible . . .": Carl Sagan, et al., *Murmurs of Earth*, pages 235–36.

282 evidence of planets around other nearby stars: The *Extrasolar Planets Encyclopedia*, online at exoplanet.eu/catalog, contains lists, plots, and links to the now more than 1,800 planets discovered around nearby stars that are (mostly) like our sun, via a variety of ground-based and space-based methods.

285 *Kepler* has found more than 1,000 planets: See the updated tally, and lots more details, on the *Kepler* mission's website at kepler.nasa.gov.

287 never wrote up our results: Trivia fans can, however, find the Biostronomy Symposium paper that Bill Borucki and I presented back in 1993, "Characteristics of Transits by Earth-Sized Planets in Binary Star Systems," in *Progress in the Search for Extraterrestrial Life*, ed. Seth Shostak (Astronomical Society of the Pacific Conference Series, no. 74, 1995): 165–72 (online at http://adsabs.harvard.edu/full/1995ASPC... 74..165B).

289 will remain essentially pristine: Sagan et al., *Murmurs of Earth*, pages 233–34.

291 "Even as they are celebrated . . .": Stephen Pyne, "Voyager: A Tribute," The Planetary Society blog, posted September 25, 2013 (online at planetary.org/blogs/guest-blogs/2013/20130920-voyager-a-tribute.html).

Postscript: NewSpace

293 controlling about thirty different spacecraft: For a complete listing of the currently active spacecraft exploring our solar system today, see en.wikipedia.org/wiki/List_of_Solar_System_probes.

294 doled out nearly $2.5 billion: See NASA's Commercial Crew & Cargo Program Office website for more details on government support of the emerging "NewSpace" sector at http://www.nasa.gov/offices/c3po/home.

Acknowledgments

The trajectory of my life has been so full of gravity assists from so many people that I hardly know where to begin. I owe the start of my career in planetary science to Carl Sagan's *Cosmos*, and especially to *Voyager*, both of which got me hooked on the thrill of mission-related science and exploration. It seems appropriate, then, that I issue a blanket thank-you to the men and women who first dreamed up the mission in the 1960s, who built and launched the spacecraft in the 1970s, who flew them magnificently past the giant planets, and who teased scientific discovery after discovery out of their dozen or so science investigations in the 1980s, and to the people who still operate them and enable us to communicate with them today, at the boundary of where the solar and interstellar winds mingle. Although I was only directly influenced by a few of the many thousands of people who brought these missions to life, I was indirectly influenced enormously by your integrated effort. I wish I could have met and talked with all of you. But at least I can see you all there, looking up and smiling at *Voyager* as part of the Pale Blue Dot.

Among the few *Voyager* team members whom I do mention specifically in this book, I want to specifically thank the following colleagues for generously giving their time for e-mails, phone calls, reviews, fact-checking, and/or in-person interviews: Suzy Dodd, Gary Flandro, Heidi Hammel, Candy Hansen, Ann Harch, Andy Ingersoll, Torrence Johnson, Charley Kohlhase, Jon Lomberg, Jamie Sue Rankin, Larry Soderblom, Linda Spilker, Ed Stone, Rich Terrile, and Randii Wessen. I want to specifically call out *Voyager* Project Scientist Ed Stone's incredible generosity of time and his enthusiasm for my attempts to try to capture many of the personal thoughts and reactions that he and his team had, and the challenges that they faced, during this decades-long adventure of a lifetime.

I would also like to thank the many friends and mentors who helped me get from small-town Rhode Island to the azure-blue shores of Neptune (and beyond) along with the *Voyagers*. My Coventry High School chemistry teacher, Dr. Barry Manley, helped set me and my best friend, Bobby Thompson—another amazing source of consistent support and camaraderie—on lifelong careers in science, partly by simply asking, "Why don't you go look that up?" Thank you to Mark Allen at Caltech for giving a freshman a fun research project to work on, even though we were supposed to be concentrating on our classes; and I owe a huge debt of gratitude to my pal and mentor the late Ed Danielson for teaching me some of the arcane art of image processing and for helping me get a fly-on-the-wall glimpse of the inner sanctum of planetary exploration—the science operations rooms of Building 264 during the Uranus flyby. And thank you, Fraser Fanale, one of my PhD committee members from Hawaii, for giving me a ticket to the same inner sanctum for the Neptune flyby three and a half years later. You were all enablers, but the good kind.

I have enjoyed working with Stephen Morrow at Dutton, Penguin Random House as much on this project as I did on *Postcards from Mars*. Thank you, Stephen, for sharing my vision of robotic exploration as really being human exploration, and driven by strong human emotions and frailties, at that. Thank you also to Michael Bourret at Dystel & Goderich for your constant support and encouragement of my work, especially this, my first attempt at trying to tell a nerdy space story without so many pictures!

Finally, I want to thank my family for their never-ending support in my journey outward, to California, then to Hawaii, and then onward, among the planets. Don't worry, I still actively seek out grinders, cabinets, bubblers, and Del's, even on other worlds. Special thanks are also due to my daughter Erin for helping with some of the voice transcriptions of my interviews, and to my wonderful, beautiful friend Jordana Blacksberg for her love, support, encouragement, editing, and spectacular cooking. I am loving this interstellar journey that we are now on.

Index

NOTE: Page numbers in *italics* refer to illustrations.

Acta Astronautica, 46
Adams, John Couch, 192–194, 211
Adrastea, 128
Alchemy, 274
Aldrin, Buzz, 8
Allen, Mark, 27–28, 30
Amalthea, 128
Ames Research Center, 135,
    283, 286
Anders, William, 227–228
Antennas
    DSN, 63, 138, 199, 200, 255
    Galileo, 129
    Voyager, 56, 57, 236, 255, 276
Apollo, 8–9, 48
Apollo 8, 227–229
Apollo 17, 229

Arago, François, 193
Archimedes, 225
Ariel, 181, 243
Armstrong, Neil, 8
Astrology, 163, 195
Atlas Intercontinental Ballistic
    Missile, 62
Atmosphere
    Earth, 139
    Mars, 285
    Neptune, 200, 205–207
    Saturn, 134, 139–141
    Triton, 215–216
    Uranus, 177, 188–189
    Venus, 285
Attitude and Articulation Control
    Subsystem, 52

313

LIFE magazine, 227, 229
Lomberg, Jon, 79, 82–84, 85, 87, 90, 91, 95, 98–99, 219–220, 282
Lopes, Rosaly, 120
Lovell, Jim, 227, 228
Lunar Orbiter I, 226–227

Magellan, 51, 195
Magnetic fields
   Earth, 171
   Ganymede, 126
   heliosphere-interstellar space boundary, 246, 249, 254, 256, 257, 259, 261, 262–264, 266, 267, 269, 271
   Jupiter, 54, 73, 110, 130, 178
   measurement tools, 16
   Neptune, 200, 209–210
   Saturn, 73, 135, 178
   Stone's research, 16–17
   Uranus, 37, 170–171, 177–180, 185
Magnetometers, 16
Malin Space Science Systems, 39
Mariner 2, 50
Mariner 4, 50
Mariner 5, 50
Mariner 6, 50
Mariner 7, 50
Mariner 9, 50
Mariner 10, 42, 47, 50
Mariner 11, 61
Mariner 12, 61
Mariner Jupiter Saturn '77 (MJS-77), 48, 49–61
Marley, Bob, 95
Mars
   atmosphere, 285
   Curiosity rover mission, 10, 26, 84
   life on, 131
   Mariner flyby, 47

Opportunity rover mission, 10, 19, 26, 84, 230, 238–239
Spirit rover mission, 10, 19, 26, 84, 230, 238–239
Viking program, 9–10, 12, 62, 157, 183
Mars Global Surveyor mission, 39
Mars Observer spacecraft, 39
Mars Orbiter Camera (MOC), 39
Mars Pathfinder spacecraft, 10
Martin Marietta Corporation, 62
Mathematicians, 192–193
Mauna Kea Observatories, 58–59, 186, 188, 195, 205
Maxwell, James Clerk, 135
McComas, David, 266
Medici, Cosimo de', II, 166
Mercury
   Galileo flyby, 237
   Mariner flyby, 42, 47, 50
   MESSENGER flyby, 237
   orbiters, 26
   planet designation, 243
   transits in front of sun, 283
   visibility from earth, 239
   Voyager photographs, 235–236
Messages sent into space, 71–99
   "Cosmic Call," 81
   introduction, 71–73
   New Horizons, 97–99
   Pioneer plaques, 73–77
   Voyager Golden Record, 77–97
MESSENGER, 237
Methane, 139, 140, 141, 146, 177, 203, 215
Metis, 128
Milky Way galaxy, 224, 251, 290–291
Miller, Stanley, 139
Mimas, 142, 150, 166, 180
Miranda, 37, 39, 172, 177, 181–183